FASCINATING ARCHITECTURE

World Public Architecture
世界最新公共建筑

高迪国际出版（香港）有限公司 编

大连理工大学出版社

Dalian University of Technology Press

图书在版编目(CIP)数据

世界最新公共建筑：汉英对照 / 高迪国际出版（香港）有限公司编. — 大连：大连理工大学出版社，2011.3

ISBN 978-7-5611-6050-3

Ⅰ.①世… Ⅱ.①高… Ⅲ.①公共建筑—建筑设计—作品集—世界—现代 Ⅳ.①TU242

中国版本图书馆CIP数据核字（2011）第022964号

出版发行：大连理工大学出版社
　　　　　　（地址：大连市软件园路80号　邮编：116023）
印　　刷：利丰雅高印刷（深圳）有限公司
幅面尺寸：246mm×290mm
印　　张：23
插　　页：4
出版时间：2011年3月第1版
印刷时间：2011年3月第1次印刷
责任编辑：袁　斌
责任校对：李　楠
封面设计：吴倩平

ISBN 978-7-5611-6050-3
定　　价：298.00元

电　话：0411-84708842
传　真：0411-84701466
邮　购：0411-84703636
E-mail: designbooks_dutp@yahoo.cn
URL: http://www.dutp.cn

如有质量问题请联系出版中心：（0411）84709246　84709043

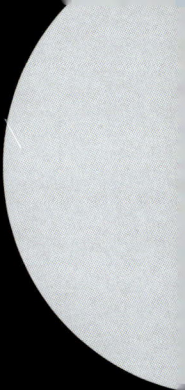

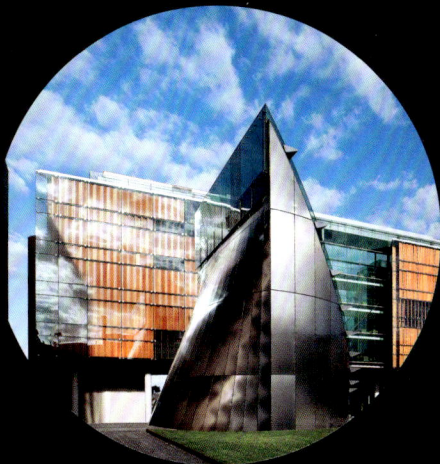

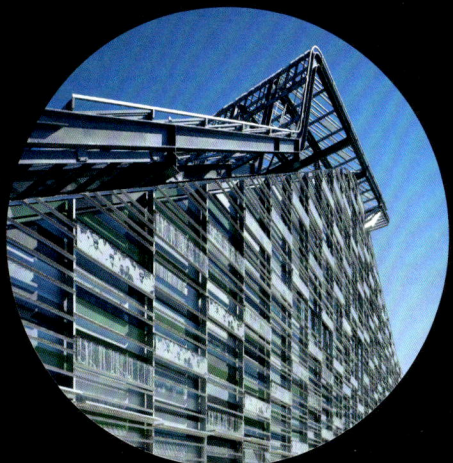

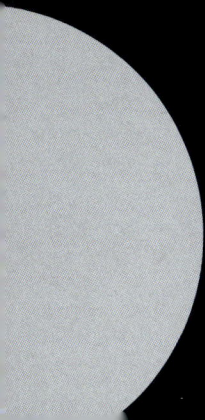

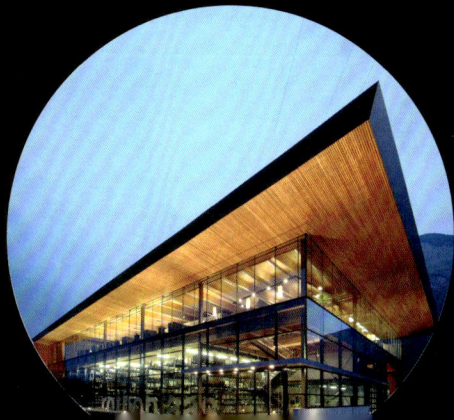

FOREWORD 序言

Antonio Vaillo + Juan Luis Irigaray
VAILLO+IRIGARAY ARCHITECTS

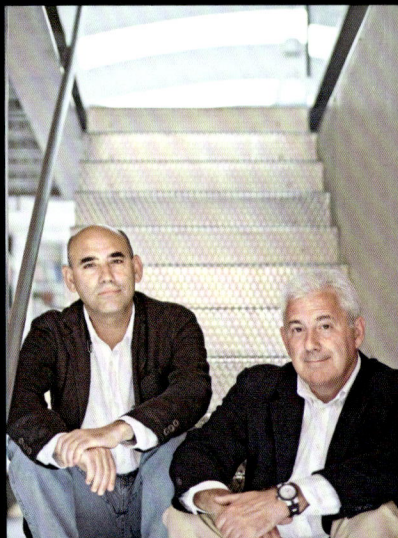

Multifocality offers solutions to every raised problem, and sets up a functional hierarchy as well as a wide capacity for shape synthesis. Constant collaboration with every agent of the complex process provides a global vision and the appropriate solution, which is also enhanced by different outlooks.

Only from the very definite is feasible to offer universal solutions. It is thus that our working method consists of the resolution of the problems raised by the project itself. The work material is, as it has been explained, the prime mover, the basis, the project concept and the essential raw material.

Our work system is based on our relationship with the client and the interplay between all intervening agents. That is precisely what shapes our projects: our project work material is the project itself, as well as its specific circumstances: location –weather and geography, place culture, client, public demand, economic data, expectations.

We fell at ease working and operating with those materials that frame the take-off point of the project. Our work approach does not "impose" any solution to the architectural project. On the contrary, the points of departure are the ones that shape the final project in its own manner, so that any other place, any other client or any other circumstances, lead to a different proposal.

VAILLO+IRIGARAY ARCHITECTS is interested in taking part in this contest to propose innovative solutions for new "architectural hybrids", capable of containing new uses and programmes: art, dwelling, retailing, offices… as well as setting flexible buildings, ready to host new deployments permanently providing a "living organism", capable to adapt easily to the wind of change. We have a history of materialising complex hybrid buildings with great functional, formal, iconographical efficacy.

Essentially, our way of acting relies on the integration of complex, diverse programmes into efficient, logical, adaptable, poetical architectural structures; we try to join functional logic and shape poetry together, searching, as begetter concepts, physical, social, cultural relationship with the environment and its own distinctive soul that makes it unique.

Multifocality为提出的每一个问题提供解决办法，并设置了一个在功能上层次分明涵盖广泛的综合体。它与每一个复杂个体的不断合作，结合它变化的外观一起提供了一个具有全球视野并合适的解决方案。

唯一非常确定的是，提供普遍的解决方法是可行的。因此我们的工作方法包含由项目本身引起的问题的决议。工作材料正如它已经解释过的，是原动力，是基础，以及项目概念和本质上未加工的原材料。

我们的工作系统是建立在我们和客户及相互影响的所有中介的关系上，这恰恰塑造了我们项目的内容：我们的项目工作材料是项目本身，也是它特定的环境：地点——天气和地形、地方文化、客户、公众需求、经济数据、期望值。

对于制定项目分支点的材料，我们认为操作起来很容易。我们的工作方法是不强加任何的解决方法在建筑项目上。相反地，出发点是那些根据自身习惯塑造的最终的项目，所以在任何其他地方，其他客户，或者其他的情况下，都会导致不同的提案的出现。

为了对"混合建筑"的革新的解决方案提出建议，VAILLO+IRIGARAY建筑事务所对参加这个比赛很有兴趣，因为比赛包含新的用途和项目：美术馆、住所、零售、办公，以及灵活性的建筑，准备好为新的发展永久地提供"活体"，具有能适应任何风云变幻的能力。我们有带有极好功能性的、正式的、具有肖像功效的、混合的综合设施建筑悠久历史。

本质上，我们的行为方式是依靠综合设施的集成，把多种多样的项目变成有效率的、符合逻辑的、适应性强的、理想化了的总体结构。我们尝试加入功能逻辑和塑造意境一起，作为生产概念、物理的、社会的和文化的关系，结合环境因素，以及它本来就与众不同的品质一起来让它变得独一无二。

Endo Shuhei

Endo Shuhei Architect Institute

Endo Shuhei today is groping for plausible ultra-modern architecture that can open up new possibilities by overcoming the self-imposed limitations of modernism in architecture that resulted from pursuing uniformity to excessive degrees, while taking advantage of its possibilities and effectiveness to the fullest extent.

In modernism, architects broke down the whole of architecture into such elements as post, beam, roof, and wall, and reassembled these elements again. Limitations inherent to this kind of "composition" may be regarded as a major reason why modernism in architecture never gained real richness in its character. Therefore, this architect has experimented with non-compositional methods in two types of architecture. The first type was based upon questioning anew a fundamental premise of architecture, namely complete separation of interior from exterior. This type of architecture consisting of open spaces has been named, to use a neologism, "Halftecture".

In the other type, architectural spaces were created with a single element, that is, continuous strips of plate encompassing both roof and wall, not with the compositional elements obtained by breaking down the whole of architecture into its elements. This was called, in a similar vein, "Rooftecture".

These buildings are concrete examples of an attempt at realizing possibilities of ultra-modern architecture. This concept had been realized in small-scale buildings such as a parking building for bicycles, a public lavatory, and an unmanned railway station.

These "Halftecture" structures encompassing ambiguous spaces were built with continuous, simple strips of steel plate only. This type of building points to a new possibility of building architectural structures. Many "Halftecture" designs have been realized with corrugated steel sheeting. This material is an industrial product manufactured by applying a wavy form to steel sheeting. These plates have structural strength, and are durable and recyclable because of their galvanized surface. They are well suited for assembling a structure on-site with bolts and nuts since they are produced in sheet form with standardized dimensions.

"The unspecified number of people experience the public building, and it is a valuable place to share the experience. In other words the times are made by this shared experience. In addition, it will be the role that every building takes in future. Therefore it is important I realize space offering a glimpse of the future to people, and to show at the same time to reflect now-the present of each times not reprinting a past."-- Endo Shuhei.

由于过度追求角度的均匀性，而导致现代建筑自我强加的局限性，建筑师远藤秀平通过克服它来实现对超现代主义建筑的合理的探索，同时尽情地利用了当中的各种可能性和有效性。

在现代主义建筑中，建筑被分解成了梁、柱子、墙体、地面和天花等等各自独立的个体，然后再把它们重新组合。这些"组合性建筑"所固有的局限性被认为是现代主义建筑的特色和无法得到真正丰富的主要原因。所以，远藤秀平对两种类型的建筑用非组合性建筑的方法进行实验。第一种类型是在基于质疑的基础上，也就是对室内外空间的重新的完全解构。这种由开放的空间包含的建筑类型，运用一个新词来命名，"半建筑风格"。

对于另一种类型建筑，建筑空间是由单独的元素创造的，也就是说，由屋顶和墙体的连绵不断的钢板组成的，而不是通过打破整个建筑而获得的构件的组合。这种类型的建筑，用相似的词组，称为"屋顶筑"。

这些建筑是超现代主义建筑的可能性实现的具体的例证。这些概念已经在一些小规模的建筑上得到实现，例如一个自行车棚、一个公共厕所、一个无人售票的火车站。

这些"半建筑风格"的结构包含了模糊不清的空间，它是仅由连绵不断的钢板的简单钢条组成的。这种类型的建筑表明了建立建筑空间结构的新的可能性。很多"半建筑风格"的设计都是由波纹钢板来实现的。这些材料是由波纹钢板加工而成的工业材料。这些钢板因为其电镀的表层而具有结构强度、耐用性并可循环再用。他们是按标准规模生产的，因此配合螺旋螺母便能适用于实地的结构。

"大多数人都体验过公共建筑，这是一个可以分享经验的有价值的地方。换句话说就是通过分享经验来创造时代。此外，每个建筑都会成为被未来接纳的角色。所以对我来讲，很重要的的一点是我意识到空间为人们提供了对未来的一瞥，同时反映现在，这是对各个时代的呈现，而不是对过去的再版。"——远藤秀平

CONTENTS 目录

Corporate
办公

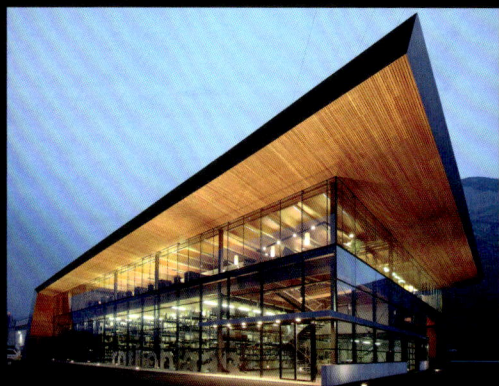

Art & Culture
艺术与文化

Commercial
商 业

Hotel
酒店

Education
教育

Entertainment & Leisure
娱乐与休闲

Other
其他

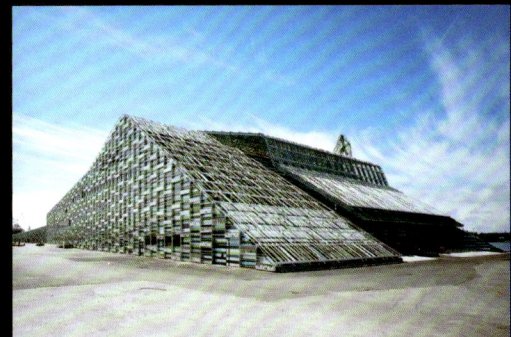

INDEX
索引

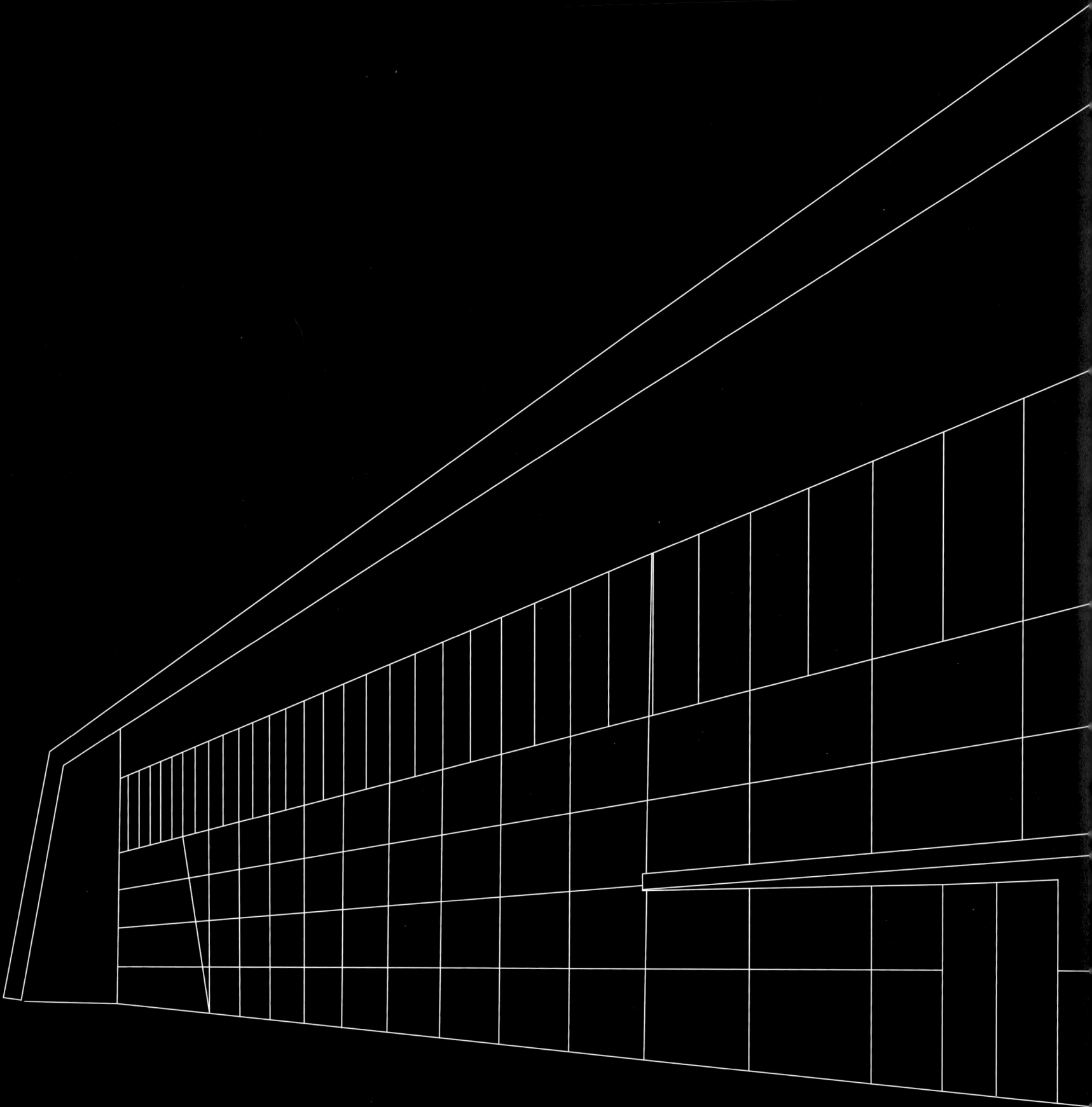

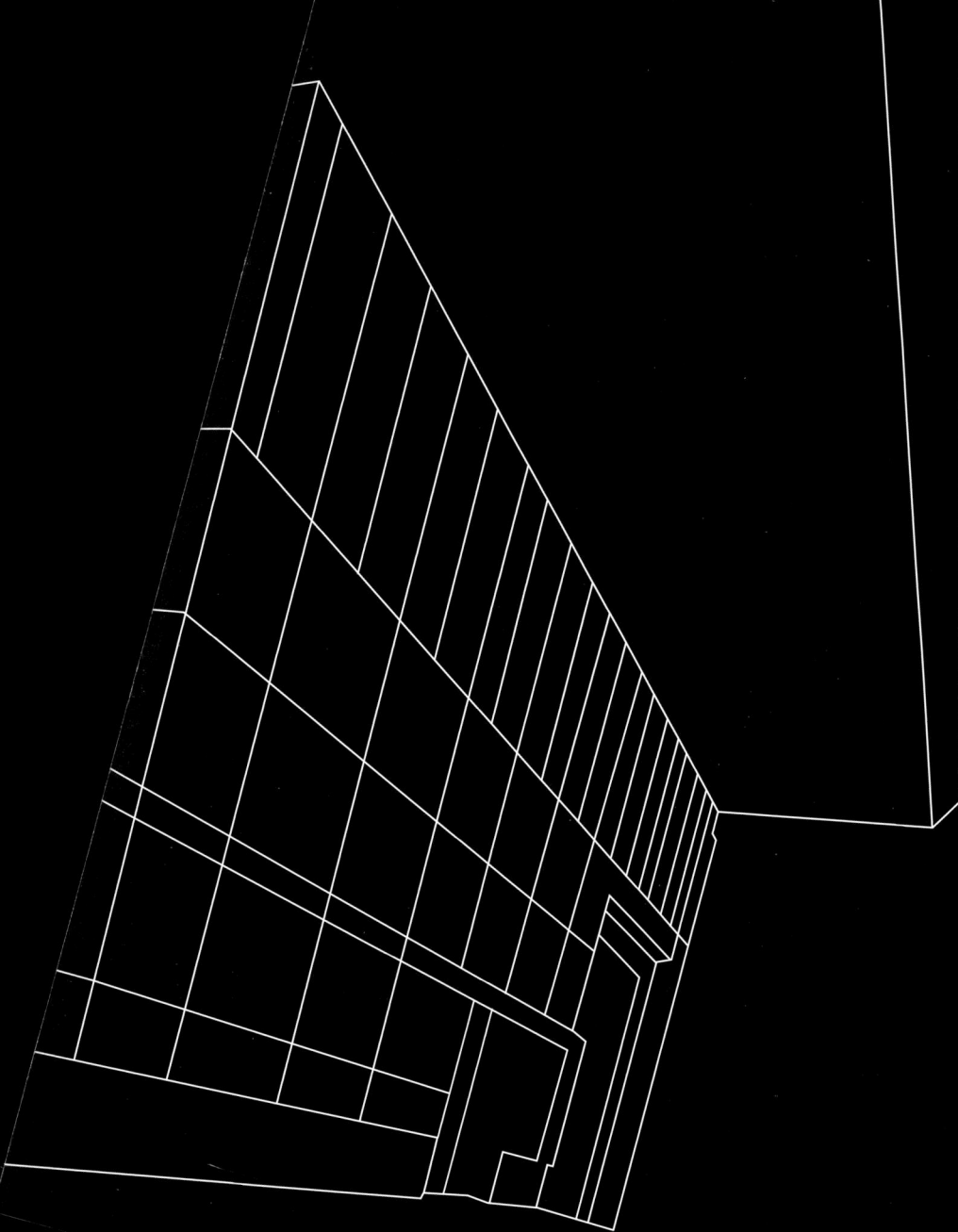

Corporate

办公

Office Silo + Containers

Architect: Antonio Vaíllo i Daniel - Juan L. Irigaray Huarte, David Eguinoa
Firm: Vaillo + Irigaray Arquitectos, SAS
Location: Tajonar
Photographer: Jose M. Cutillas, Pedro Pegenaute, Daniel Galar

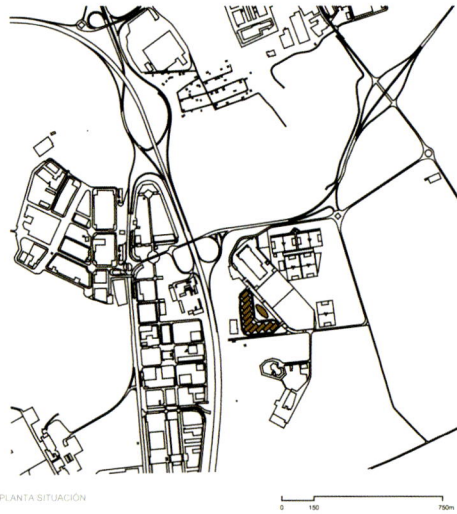

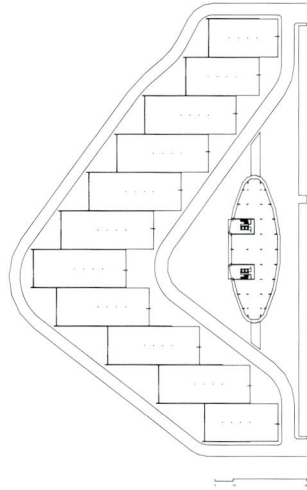

PLANTA SITUACIÓN

The area of deployment meets a non-place is located between peripheral sites and access south to Pamplona.

NAVES: There are 11 containers with a configuration in zig-zag giving the image away from a joint model, closer to industrial sites.

The construction system of vessels is based on two laws of generative: a. economy of means and constructive simplicity. b. Use of poly-functional: it has tried to resolve a construction system effectively determining the maximum potential, providing high resolution with the homogenization of high performance and minimal variations solving.

OFFICE BUILDING: The office building is made by an elliptical edification that is intended to be the counterpoint to the ships: in front of a building and pointed landscape, a taller tower, curved, looking for the minimum "friction" with the other buildings.

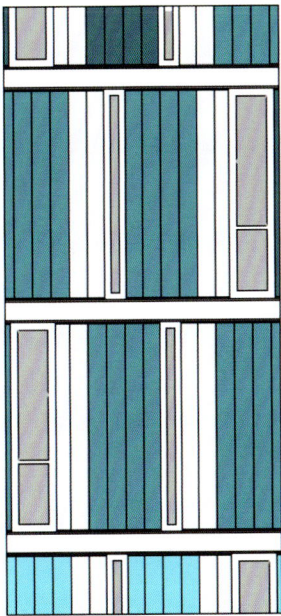

MÓDULO DE DESPIECE DE FACHADA

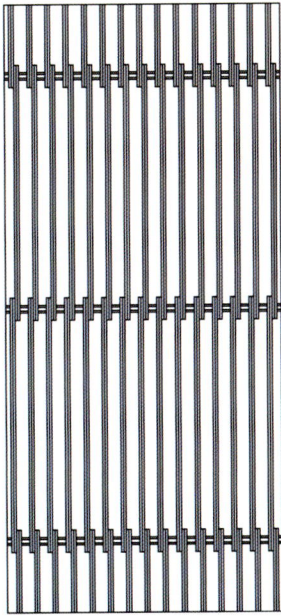

LAMAS DE FACHADA

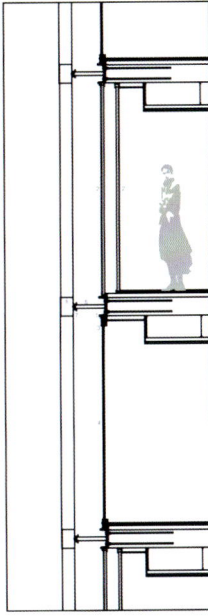

SECCIÓN FACHADA

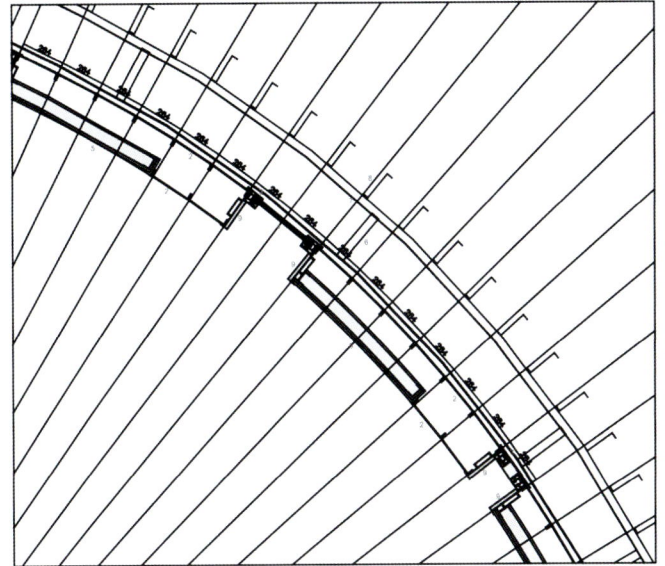

PLANTA DETALLE DE FACHADA

0   10cm   100cm

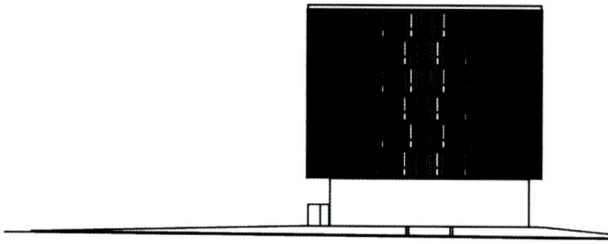

ALZADO TRANSVERSAL SUDESTE SIN LAMAS

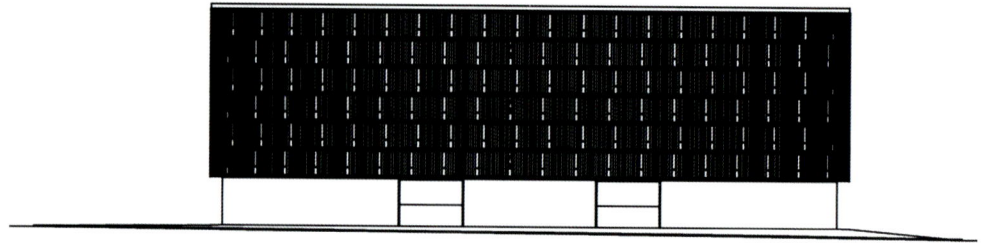

ALZADO LONGITUDINAL SUROESTE SIN LAMAS

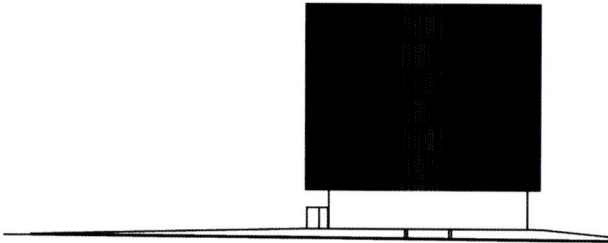

ALZADO TRANSVERSAL SUDESTE CON LAMAS

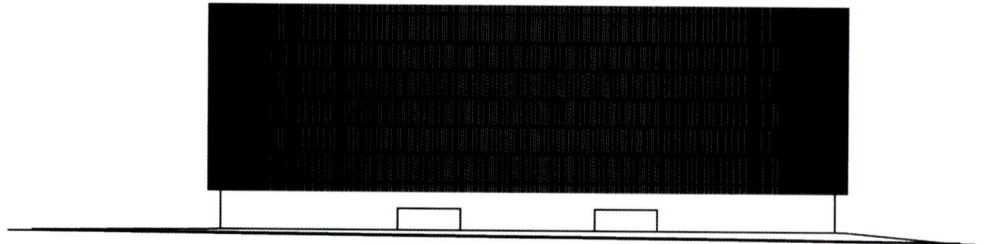

ALZADO LONGITUDINAL SUROESTE CON LAMAS

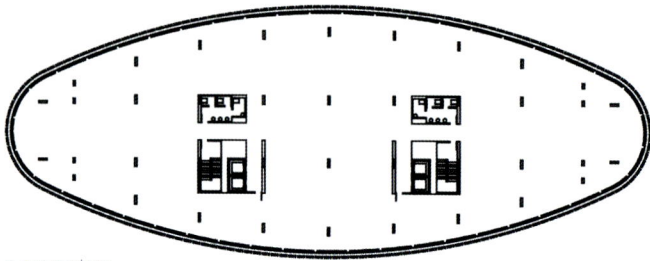

PLANTA GENÉRICA

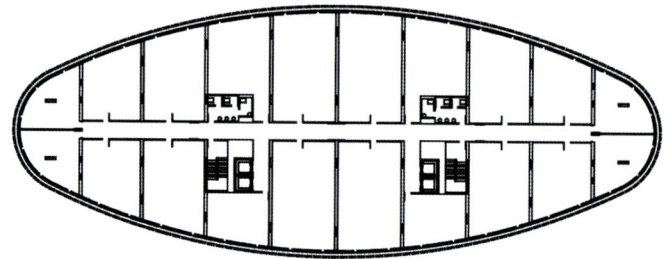

PLANTA COMPARTIMENTADA

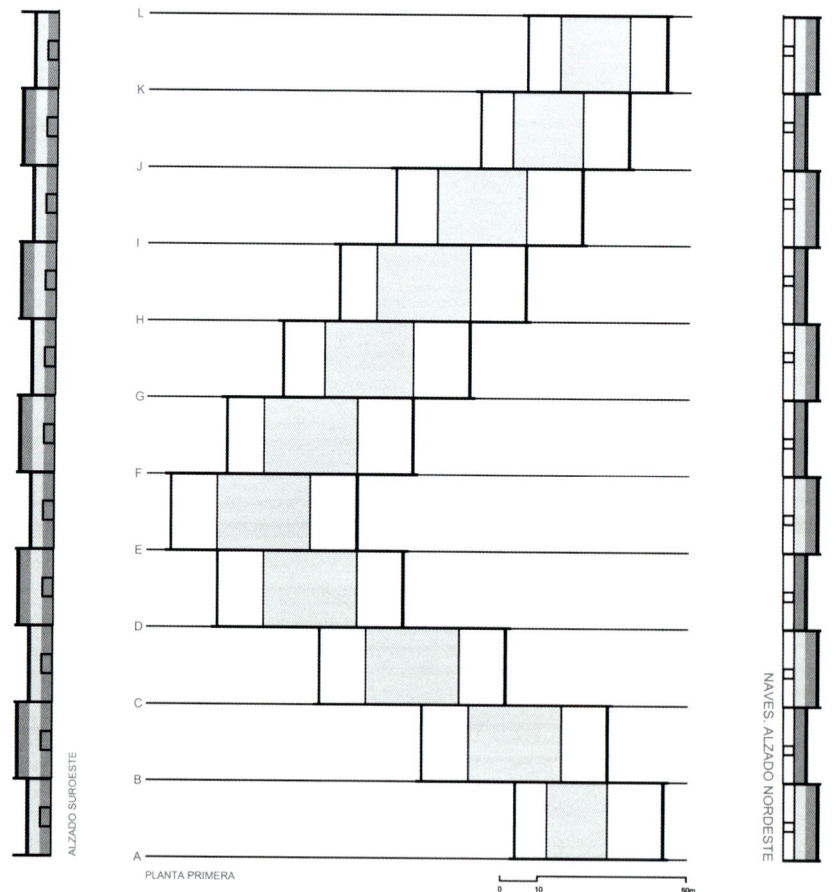

ALZADO SUROESTE

PLANTA PRIMERA

NAVES. ALZADO NORDESTE

0   10   50m

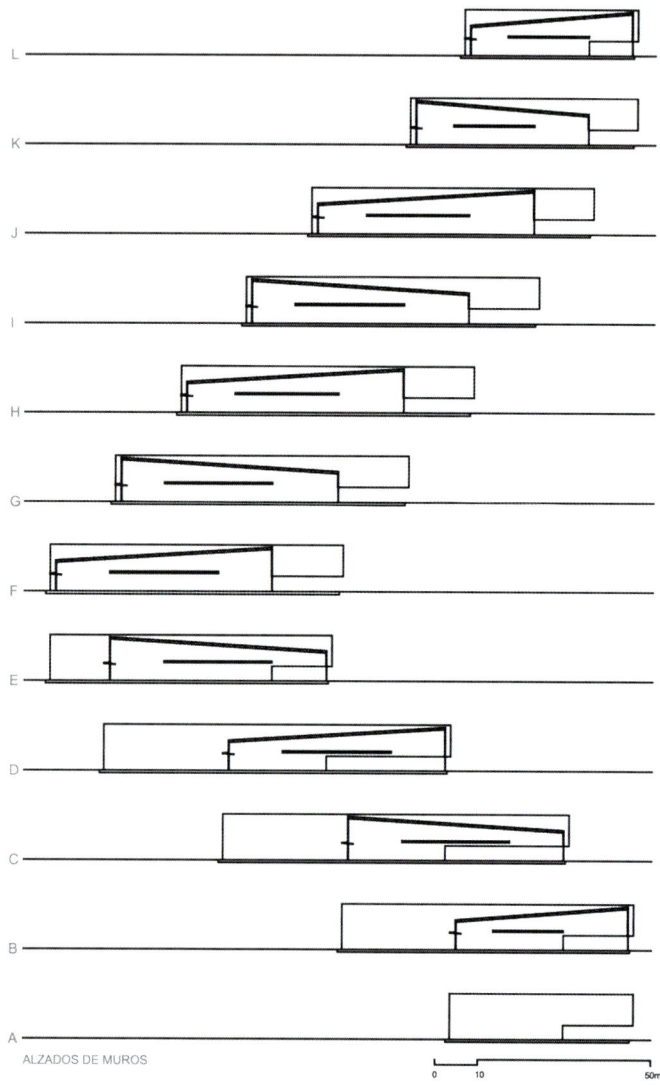

ALZADOS DE MUROS

0 10   50m

这里的布局使得所有的空间都不伸展至外围并且其南面可到达潘普洛纳。

建筑中殿：11个集装箱的蜿蜒排列结构使得建筑摆脱关节模型的形象，而让其更加类似工业建筑。

船舶造型基于两条生产规律：A.生产方法经济化及结构上的简化。B.功能多样化：它尝试解决决定最大潜力的结构系统问题，提供均化高效能及低效能的解决方法，将效能波动最小化。

办公大楼：办公大楼原本的设计构思是为了与船舶造型相呼应，设计一个高耸并弧度优美的高楼，使之成为突出的风景线的同时还能够与周遭的建筑融和。

ZMS Schwandorf Incineration Plant

Architect: Bernd Lederle, Wolfgang Heckmann, Tina Kierzek, Tilo Weber,
Katharina Schneider, Jonas Beer, Achim Zumpfe
Firm: Archimedialab
Location: Germany
Area: 38000m²

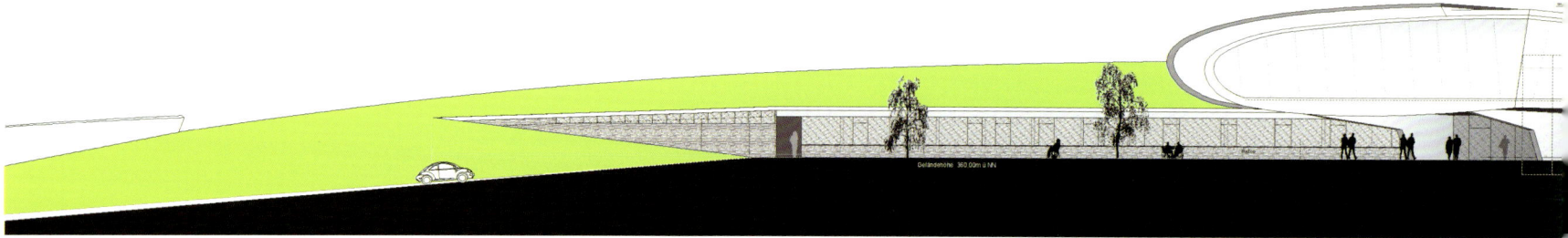

The task is to design a new administration building, reorganize the power station compound and create a new noise protection barrier of 450m long and 13m high, the central part of the noise protection wall with a 45 degree incline. The building is divided into two parts, one part simultaneously constitutes a new administration building for over 140m enclosed by green wall, the other part is the big hall that higher one level than the administration buliding with a roof uncovered by green wall.

The main structural challenges arose from the fact that we had to create an earth wall of up to 13m high that is capable of supporting not only itself at a 45 degree slope, but also accommodating a building within it.

A specially engineered and carefully balanced mixture of earths with fine aggregate and a very low percentage cement for adhesion ensures the stability of the earth wall structure, yet allows enough water to be retained to let vegetation cover the structure completely and evenly. The upper floor of the building folds out of the wall and cantilevers up to 20m over the upper level of the site. A pre-stressed concrete structure, initially supported by two curvilinear walls allows the auditorium and visitor center to hover over the site and allows panoramic views of the entire area.

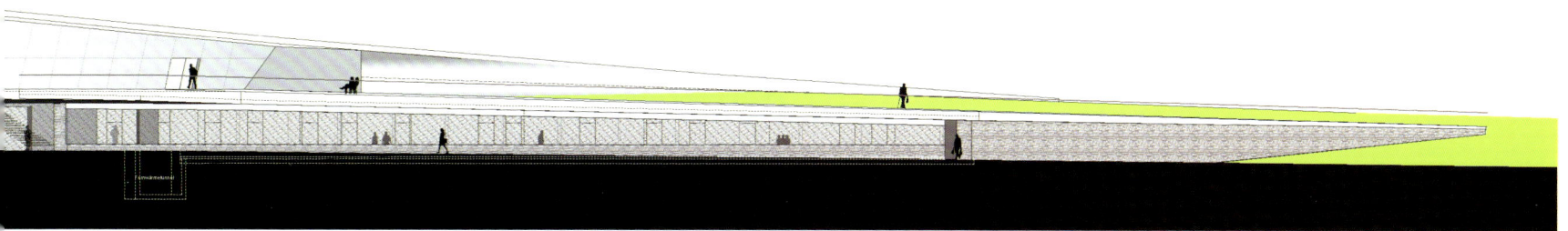

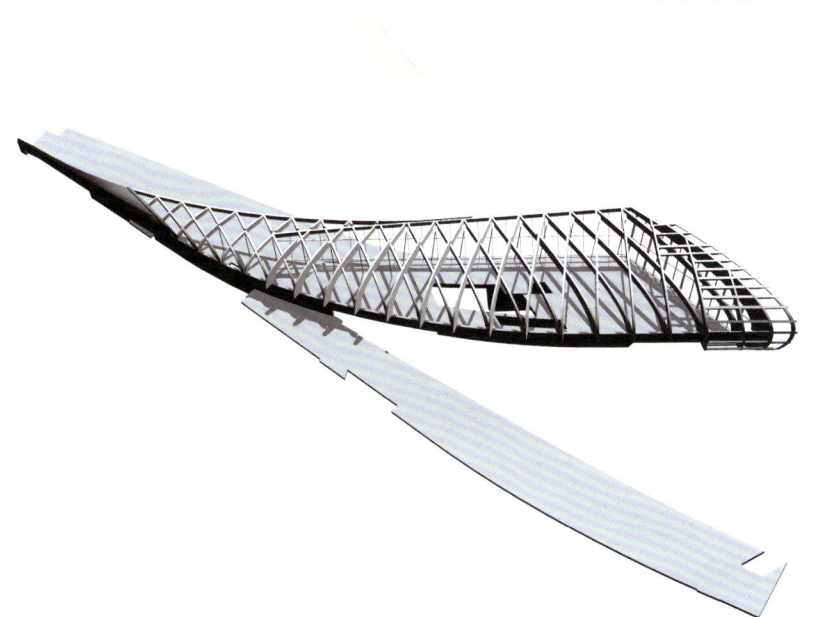

Dachaufbau 1:
a) Blecheindeckung
d) 16cm Mineralfaserdämmung
e) Dampf sperre, bituminös
f) Rauhspundschalung 2x28   bzw. 1x56mm
h) 60 mm Akustikelemente und Massivelemente zwischen den Rauten
i) 1m -0,6m Brettschichtholzkonstruktion

Dachaufbau 2:
Kunstoffabdichtung
16cm EPS-Dämmung
Dampfsperre, bituminös
Rauhspundschalung 2x28 bzw. 1x56mm
1m -0,6m Brettschichtholzkonstruktion

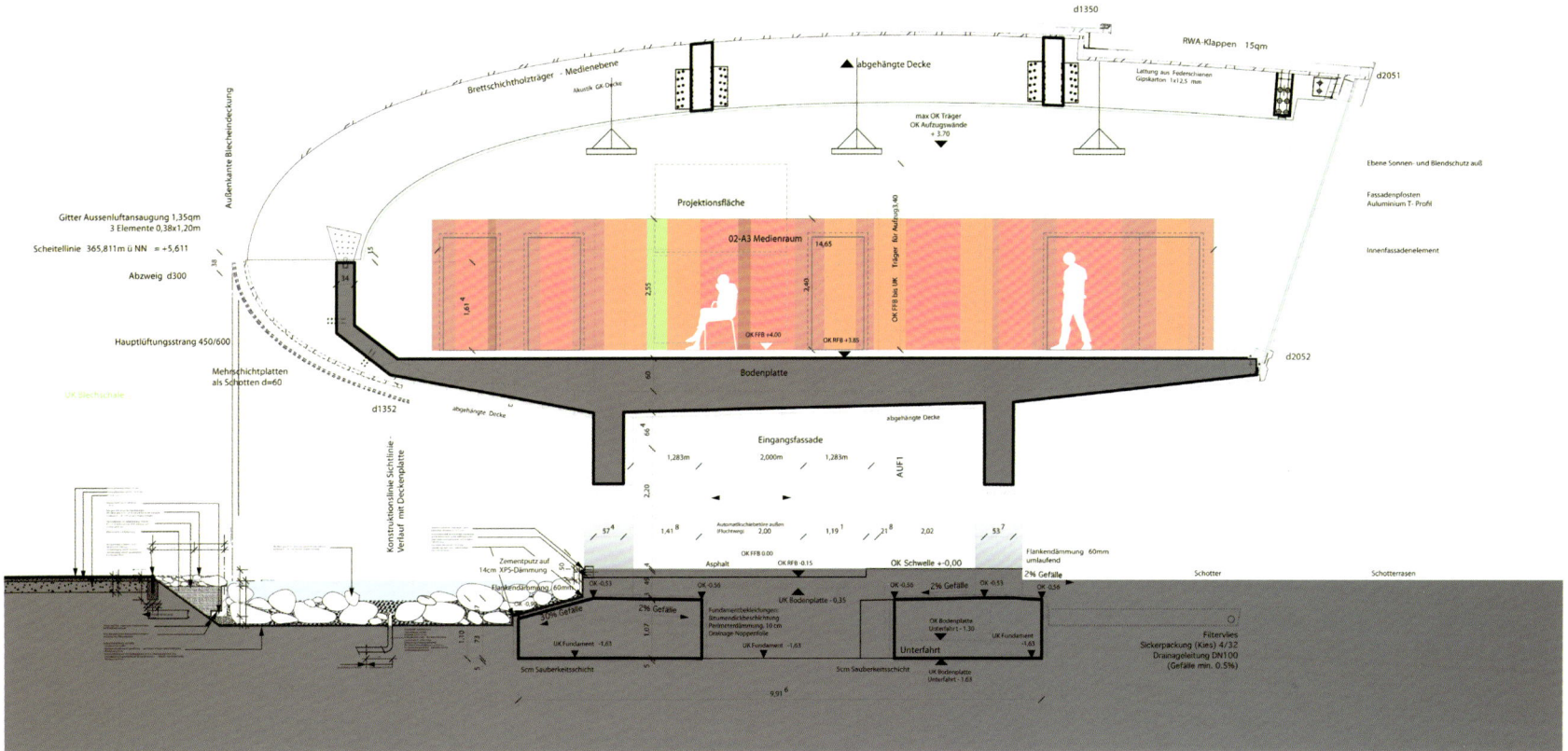

d1350

RWA-Klappen 15qm

Brettschichtholzträger - Medienebene

abgehängte Decke

Leitung aus Federnschienen
Gipskarton 1x12,5 mm

d2051

max OK Träger
OK Aufzugswände
+ 3,70

Ebene Sonnen- und Blendschutz auß

Fassadenpfosten
Aluminium T- Profil

Projektionsfläche

02-A3 Medienraum

14,65

2,55

2,40

OK FFB +4,00

OK RFB +3,85

OK FFB bis UK   Träger für Aufzug 3,40

Innenfassadenelement

Gitter Aussenluftansaugung 1,35qm
3 Elemente 0,38x1,20m

Scheitellinie 365,811m ü NN = +5,611

Abzweig d300

Hauptlüftungsstrang 450/600

Mehrschichtplatten
als Schotten d=60

Bodenplatte

d2052

Außenkante Blecheindeckung

Außenkante Blecheindeckung

abgehängte Decke

abgehängte Decke

d1352

Eingangsfassade

1,283m   2,000m   1,283m

AUF1

Konstruktionslinie Sichtlinie
Verlauf mit Deckenplatte

Zementputz auf
14cm XPS-Dämmung
Flankendämmung 60mm

Automatikschiebetüre außen
(Fluchtweg)   2,00

Asphalt

OK FFB 0,00

OK RFB -0,15

OK Schwelle +-0,00

Flankendämmung 60mm
umlaufend

2% Gefälle

Schotter

Schotterrasen

57⁴

1,41⁸

1,19¹

2,02

53⁷

OK -0,53

OK -0,36

UK Bodenplatte - 0,35

OK -0,30

OK -0,55

OK -0,59

2% Gefälle

2% Gefälle

Fundamentabdichtungen:
Bitumendickbeschichtung
Perimeterdämmung 16cm
Drainage Noppenfolie

OK Bodenplatte
Unterfahrt -1,30

Filtervlies
Sickerpackung (Kies) 4/32
Drainageleitung DN100
(Gefälle min. 0,5%)

1,07

UK Fundament -1,63

UK Fundament -1,63

Unterfahrt

5cm Sauberkeitsschicht

5cm Sauberkeitsschicht

UK Bodenplatte
Unterfahrt - 1,63

9,91⁶

024

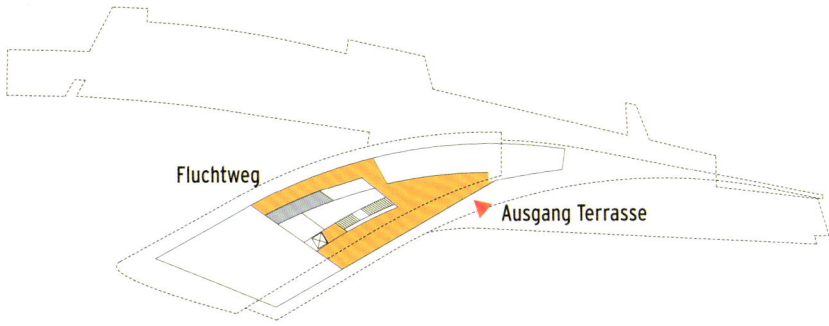
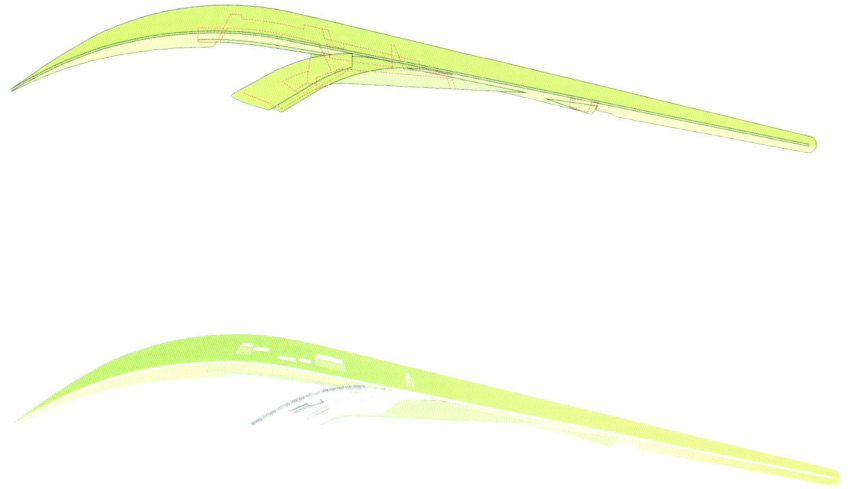

Fluchtweg

Ausgang Terrasse

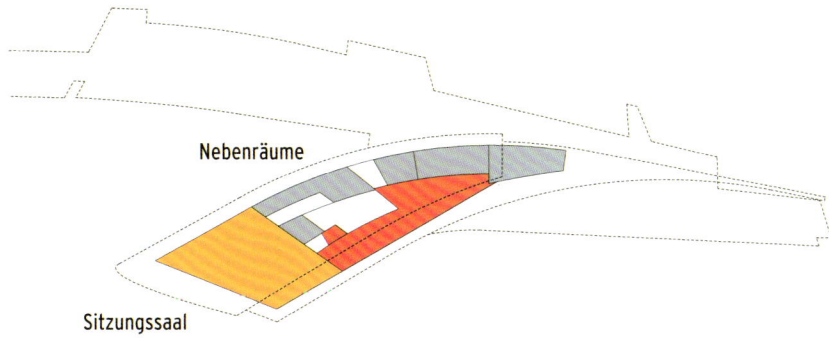
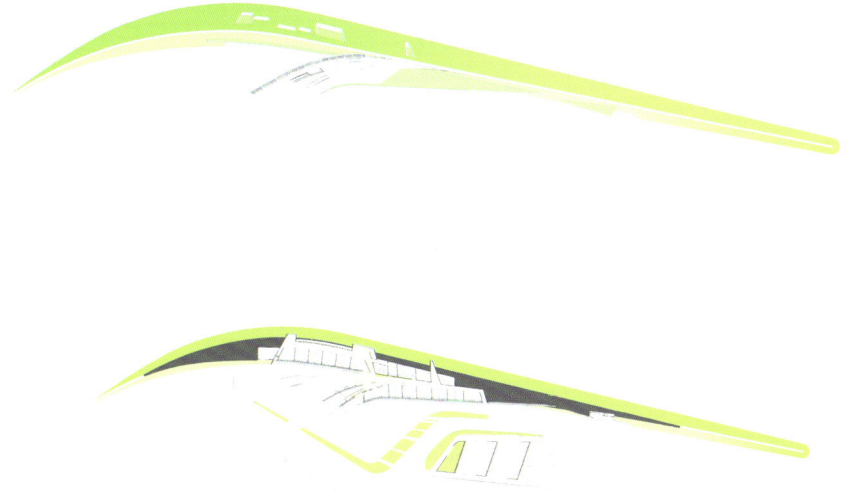

Nebenräume

Sitzungssaal

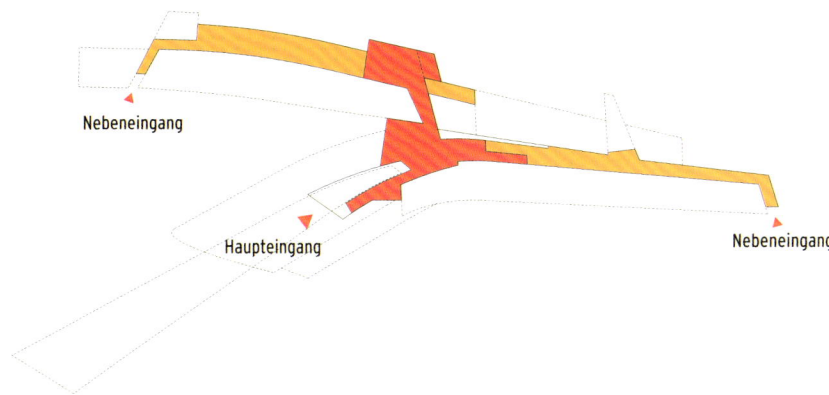
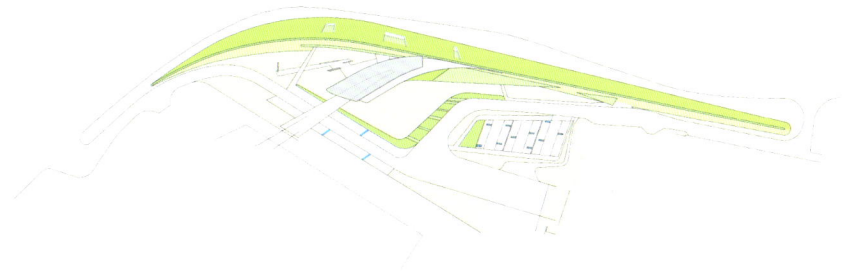

Nebeneingang

Haupteingang

Nebeneingang

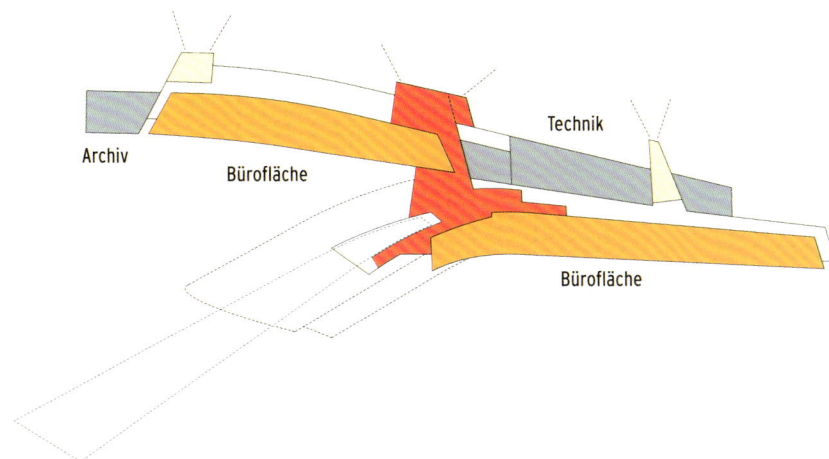
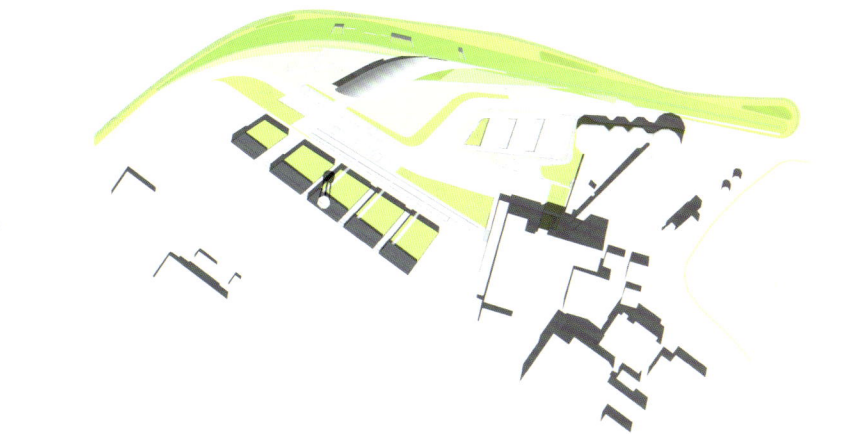

Archiv

Bürofläche

Technik

Bürofläche

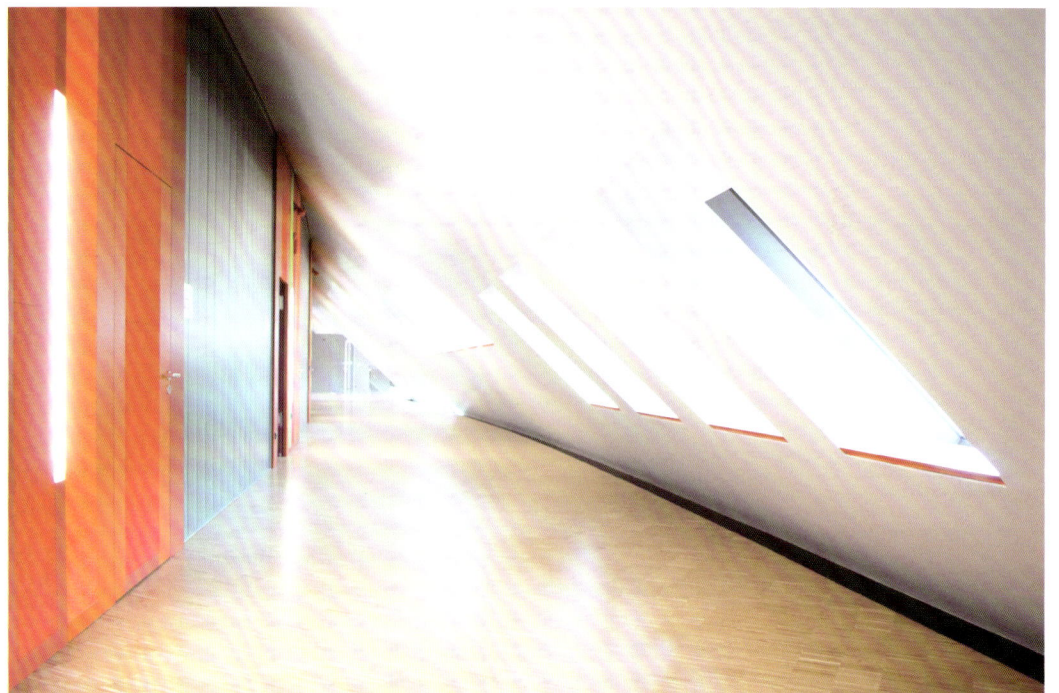

　　该项目是设计一个新的行政大楼，重新为发电厂内建筑进行布局，并建造一个新的防噪音墙。防噪音墙长450米，高13米，中央部分的倾斜角度约为45度。其中建筑分成两部分，一部分是处在绿色墙壁之下的新行政大楼所在处，长140多米；另一部分则是比行政大楼要高一层的大礼堂，屋顶并没有被绿色墙壁所覆盖。

　　建造过程中遇到的一大难题就是如何建造这个13米高的防噪音绿色墙。使它不仅能够以45度角支撑起自身，而且还能覆盖下面的建筑。

　　经过专门的研究设计，再仔细地平衡泥土、细集料和少量的黏合水泥的百分比，最终建造出这个牢固的墙壁结构，而且还能吸附部分水份，让植被能更好地生长。大礼堂所在的楼层，悬浮在空中，伸出去约20米。它是一个预应力混凝土结构，由两个径向承重臂支撑，能支撑起满人的大礼堂。

Foro Europeo

Architect: David Eguinoa, Antonio Vaíllo i Daniel - Juan L. Irigaray Huarte
Firm: Vaillo + Irigaray Arquitectos
Location: Ugarrandía, Huarte. Navarra
Photographer: José Manuel Cutillas, Antonio Vaillo

The reproductive structure of the project, part of a minimum basic unit, which likes space optimal minimum dimensions can be multiplied and can take appropriate structures for different uses. This is achieved by providing the building with great flexibility, according to the changing needs of such a program.

The building takes up the tradition of "teaching" in Europe, and takes the concept of "faculty" as a formalized structure: in this way the program is built around a courtyard.

The "clothes", which like "veils" with various rankings of transparency, respond to an abstract mimicry relationship with the natural surroundings: the subtle reflections (like those of spider webs), the ring brightness Baroque as a huge embroidery, with bare trunks in the winter or the color of the leaves in autumn, and so on.

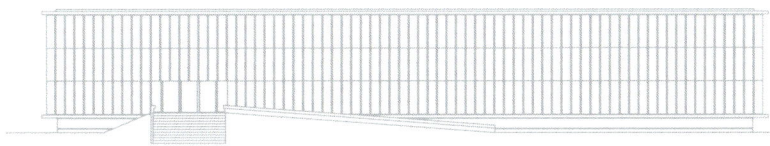

ALZADO

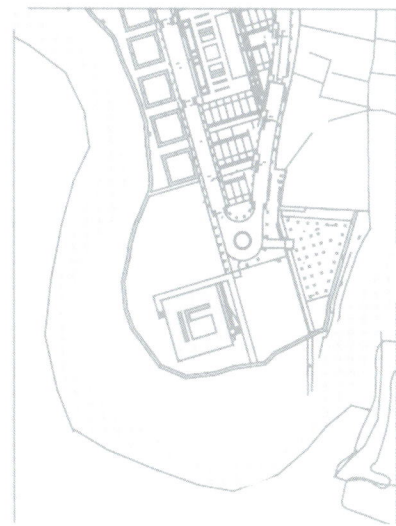

SECCION 1

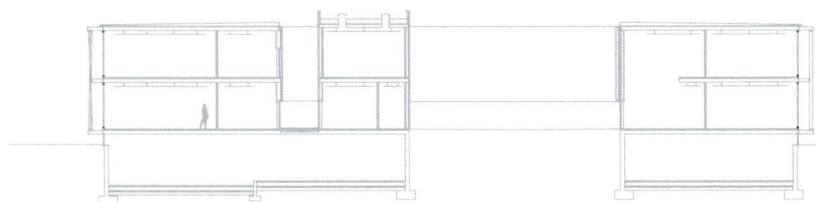

SECCION 2

0 1   5   20 m

LAMA TIPO 1

LAMA TIPO 2

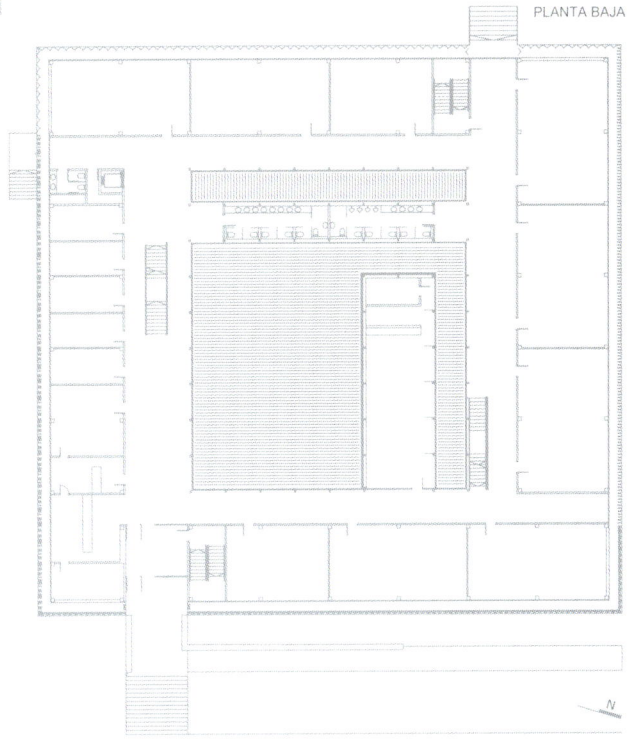

N

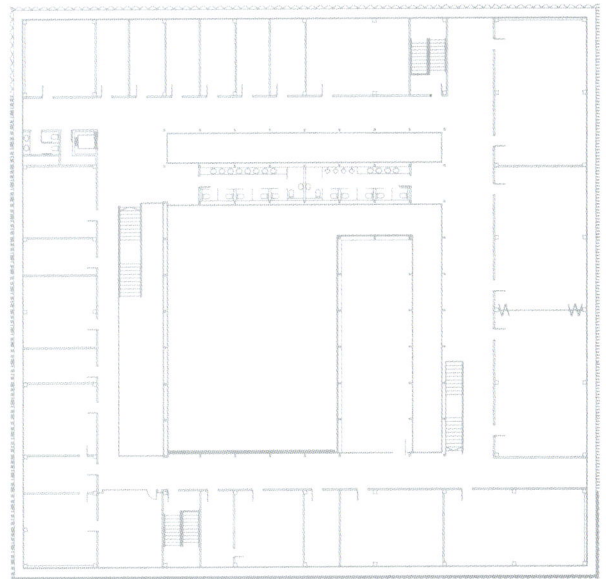

0 1   5   20 m

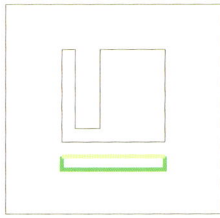

PLANTA SOTANO

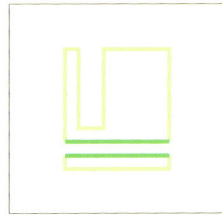

PLANTA BAJA

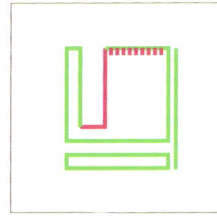

PLANTA PRIMERA

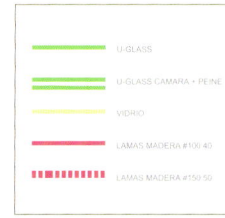

U-GLASS

U-GLASS CAMARA + PEINE

VIDRIO

LAMAS MADERA #100 40

LAMAS MADERA #150 50

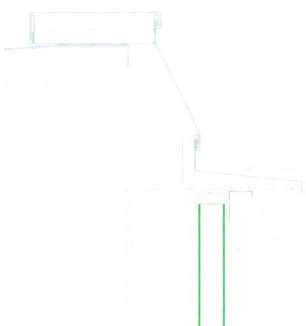

S1-A

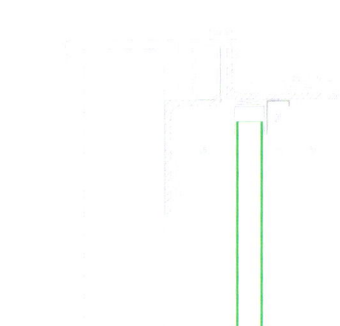

S2-A

S1-B

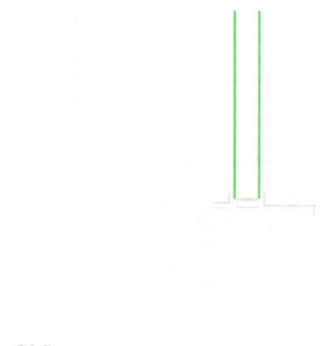

S2-B

LAMAS DE MADERA

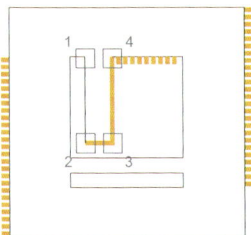

IROCO MICROLAMINADO

CORREDOR

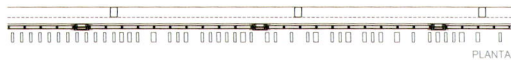

PLANTA

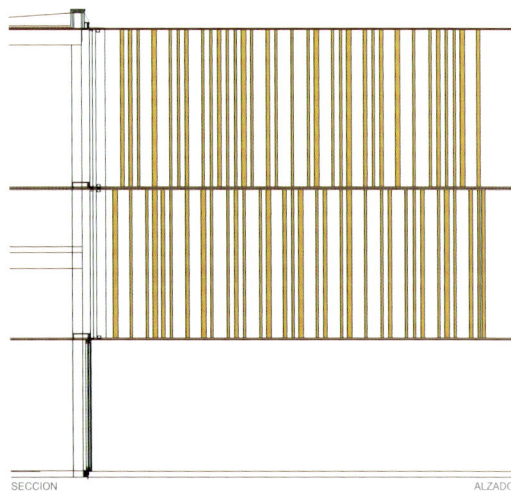

SECCION   ALZADO

BIBLIOTECA

PLANTA

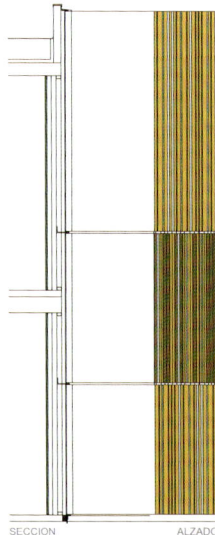

SECCION   ALZADO

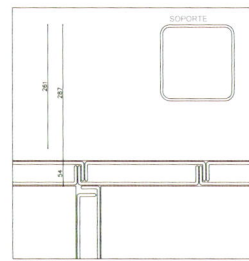

DETALLE 1

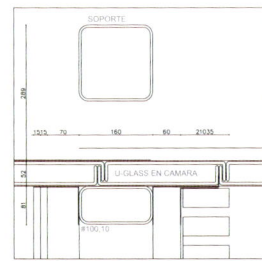

DETALLE 4

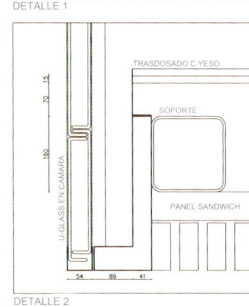

DETALLE 2

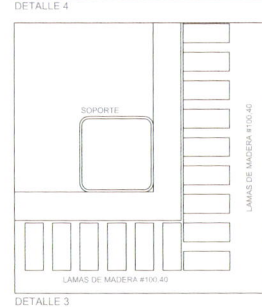

DETALLE 3

ZOOM +

ZOOM -

这个项目的重复性结构是其最小的基本单元的一部分，正如最佳的最小空间一样，可以通过叠加并采用合理的结构，来实现不同的用途。通过为建筑提供极大的灵活性，来满足这样一个项目不断变化的需求。

这座建筑吸收欧洲"教学"的传统，以"系"的概念围绕庭院建成一个正式的建筑。

"衣"有着如同"面纱"般不同的透明度，它与自然环境的抽象模拟相辉映：微妙的反光（就像那些蜘蛛网一样），巨大的巴洛克刺绣环的亮度，伴随着冬天光秃的树干或秋天落叶的颜色等等。

TransPort - transavia.com and Martinair Headquarters

Architect: Paul de Ruiter
Firm: Paul de Ruiter architects bv
Location: Piet Guillardweg, Schiphol-East
Area: 12400m²

The building's streamlined form evokes flying and gives expression to the aviation companies' identity. The building is innovative, open and public-centered. It expresses social consciousness through the use of sustainable and energy-saving technology. This has been acknowledged through the LEED Platinum assessment, which demonstrates that the building complies with international sustainability standards. It provides a pleasant working environment for the staff and is recognisable to the outside world.

The building is extremely transparent, with aluminium slats as permanent sun blinds fitted along the south side, on the inside of the curve. These sun blinds reflect the sun's heat but let plenty of daylight through. The building radiates a sense of openness, and the daylight enhances workplace quality. The latter is extremely important, as workplace quality has a significant influence on staff productivity and health. An attractive workplace lit by natural light also has a positive impact on the atmosphere at work, and on staff motivation and effort. Parking is available under the building. As the building is on pillars, there is good lighting and ventilation. Trees have been planted in the parking garage that grow up through the ceiling into the interior, lending the building a natural feel.

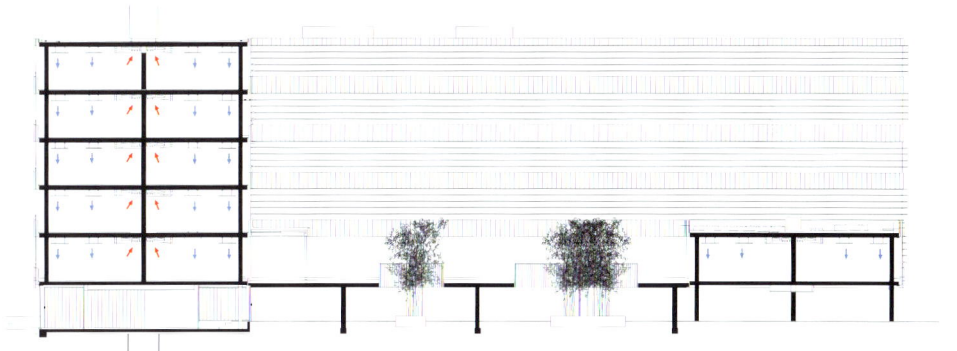

Section
Scale 1:200

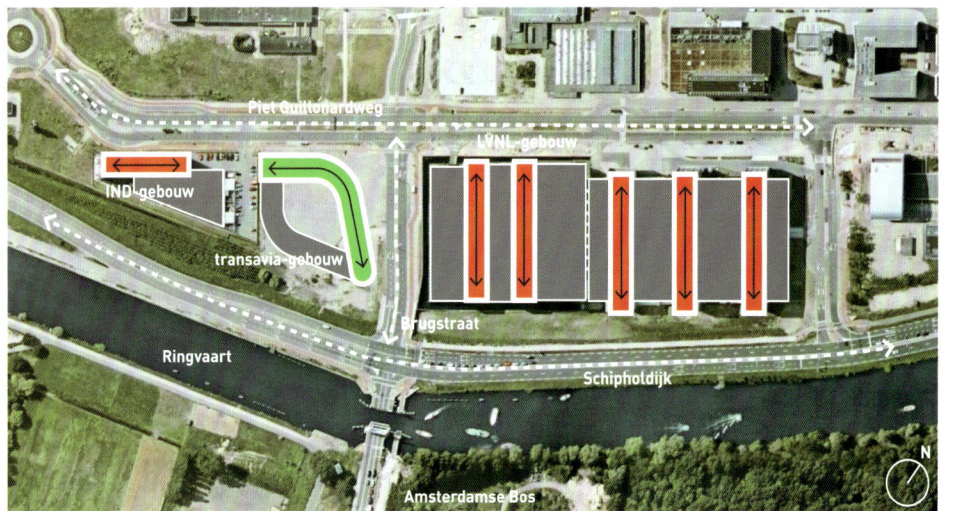

该建筑简洁的线条体现了航空企业的企业文化。这座大楼理念创新、开放、以人为本。它表达了可持续和节能的社会意识。通过了LEED白金认证的评估，达到了建设可持续性发展的国际化标准。这座建筑能为员工提供一个舒适的工作环境，外观也十分引人注目。

该建筑朝南的外墙采用铝板反射太阳的热量，并让大量太阳光通过，使得它看起来十分通透、明亮而开放。提高了员工的工作效率。工作场所的品质对于员工的生产力和健康都有重大影响。一个明亮的工作环境能让员工更加积极努力。停车场设在大楼地下层。大楼由柱子支撑，使得它既明亮又通风。停车场内种植树木，这些树穿过地下层的天花板越入大楼室内，给大楼增添了大自然的感觉。

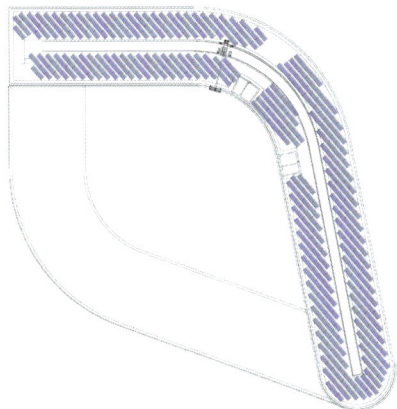

Roof TransPort
Scale 1:500

Floor 2 - 5 TransPort
Scale 1:500

1st floor TransPort
Scale 1:500

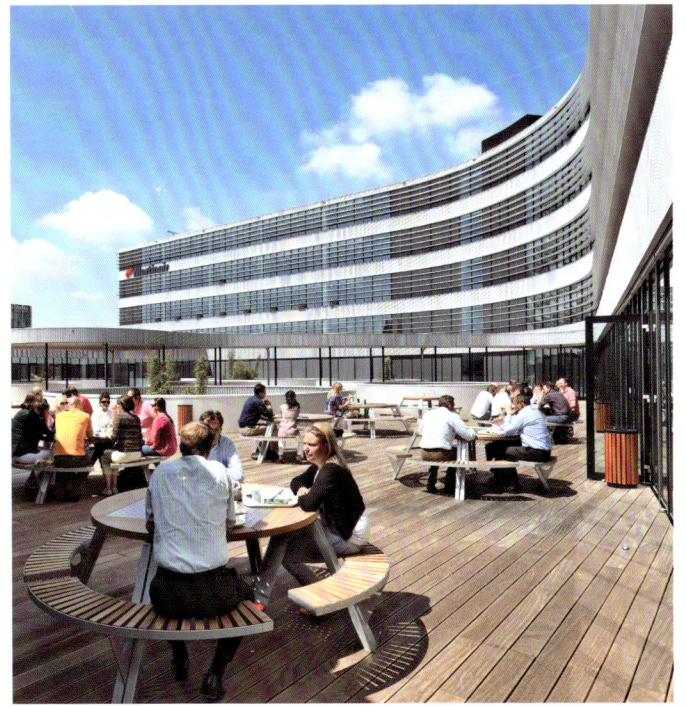

Rothoblaas limited Company

Architect: Patrik Pedó and Juri Pobitzer
Firm: Monovolume architecture + design
Location: Kurtatsch, Italy
Area: 3700m²
Photographer: Oskar Da Riz

The "rothoblaas" office is a large-scale commercial operation specializing in assembling systems and power tools for the woodworking industry. Warehouse and commissioning are situated on the ground floor whereas administration, a meeting room and a showroom can be found on the upper floor. The aim of the project was to create a compact building with a high level of recognition. The contemporany of the company becomes a corporate identity of the enterprise. This has lead to a functional, compact structural shell, provided with a glass envelope. The main building material employed is wood in order to show the own products.

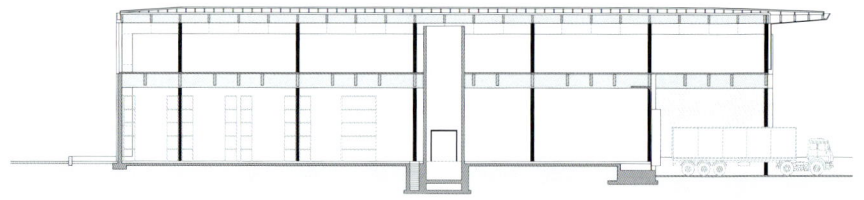

sezione b_b

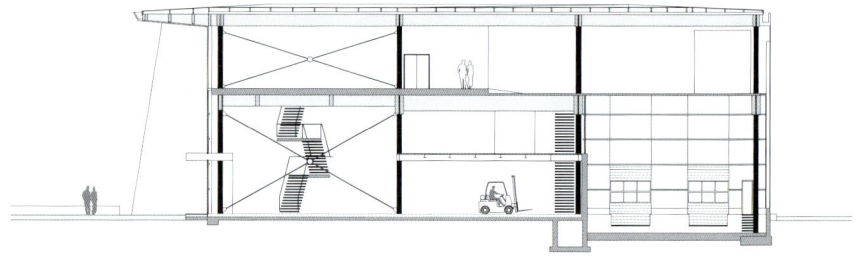

sezione c_c

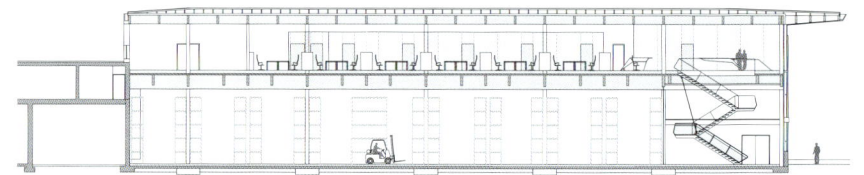

sezione a_a

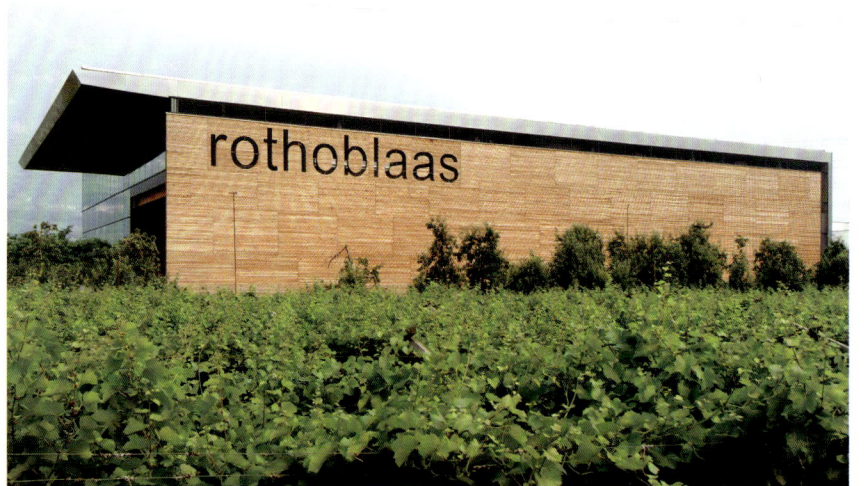

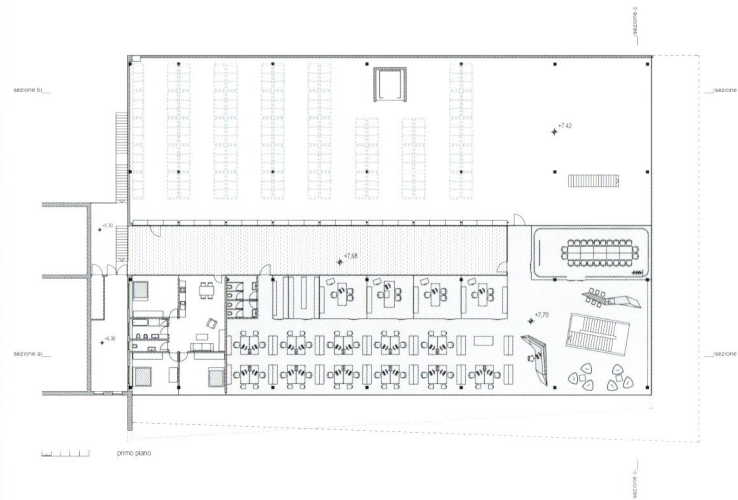

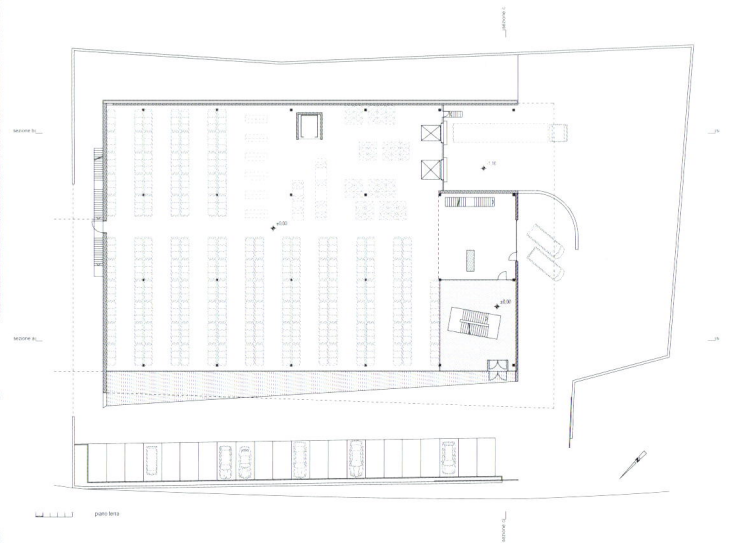

Rothoblaas办公楼内的公司是一家经营木材加工行业集成系统和大型工具的大型商业机构。仓库和调试车间位于建筑的一层，行政室、会议室、陈列室位于二层。建筑高度紧凑，具有标志性。建筑的现代感成为企业形象的标识。建筑的玻璃外壳具有功能性和紧凑感。建筑使用的另一个主要材料是木材，以显示公司的产品。

Blaas General Partnership

Architect: Patrik Pedó and Juri Pobitzer
Firm: Monovolume architecture + design
Location: Enrico Fermi Street 18, 39100 Bolzano, Italy
Area: 1250m²
Photographer: Oskar Da Riz

The company Blaas in Bolzano is specialized in electro-mechanics. In the new head office the company presents its new product range and offers repair service.

On the ground floor of the building there is the sales division, on the first floor are the exposition area and the repair shop. All administration offices are located on the second floor. The overall impression of the structure is a homogenous and closed building. Nevertheless, there exists a separation between the public and the private sector. The client can perceive this clear and formal internal division already from the outside.

The glass façade on the Northern side provides a maximum of visibility and transparency to the exhibition and sales area. The private spaces such as repair offices, stockrooms and offices have their façades exposed to the South, East and West which are protected with a sun screening system.

In order to establish an optimal relation between natural light, development and planning of spaces there has been created a luminous entrance hall in the center of the building with an inner courtyard. This green open spot permits the administrative sector of the second floor to receive ample natural light and at the same time it generates a protected, quiet recreation area for the staff.

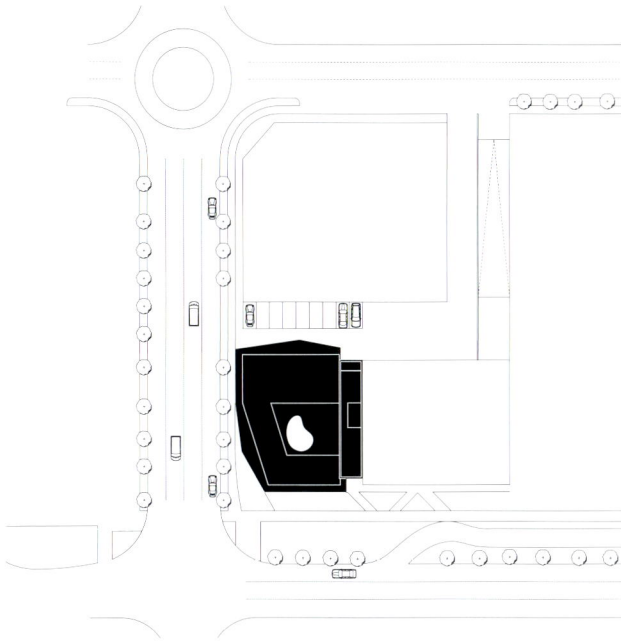

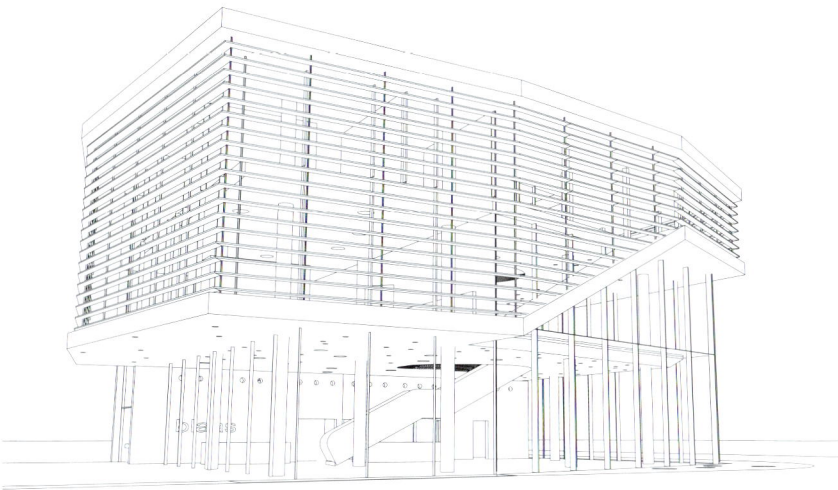

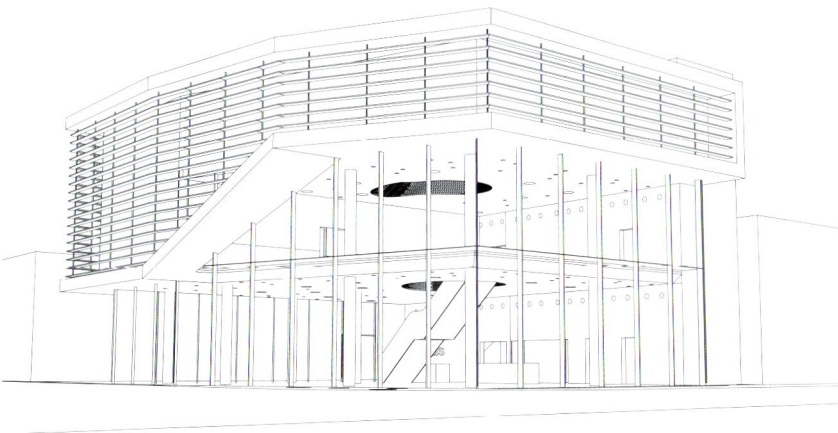

exterior

interior

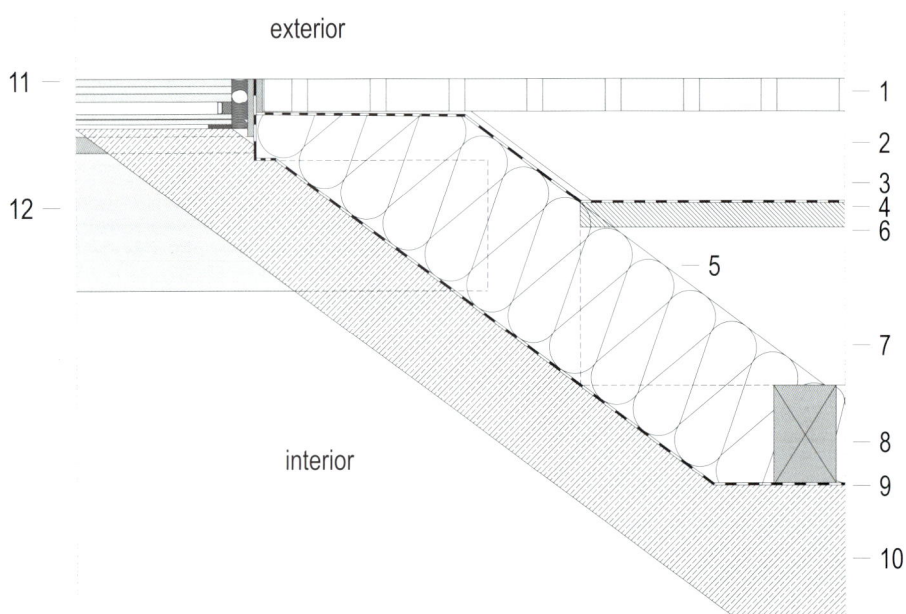

Skylight with horizontal glazing

Detail section

1 80/40 mm larch grating on bearers
2 slope square timber
3 protective separating layer
4 waterproof membrande Sarnafil G410 - 18
5 30 mm timber boarding
6 120 mm thermal insulation
7 slope square timber
8 80/120 mm wood boarding
9 vapour barrier
10 400 mm reinforced exposed concrete floor slab
11 structural glazing: laminated saftey glass of heat-strengthened glass 6 + 6 - 15 - 15 - 5 + 5 mm / ug = 1,1 W/qmK
      enamelled edge strip, black, jointed with black silicone;
12 Steel beam T 60/160/10 mm

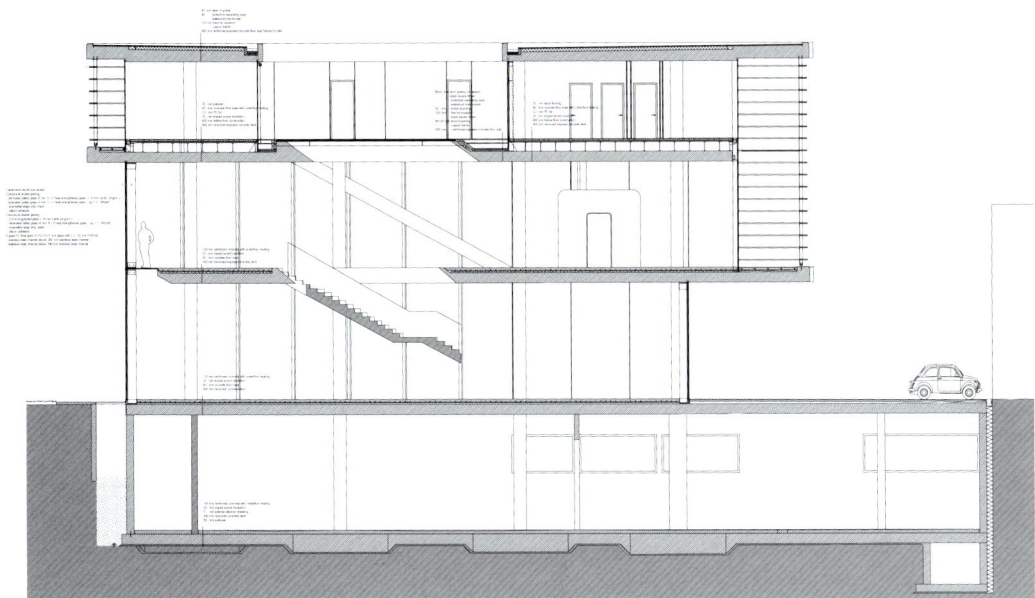
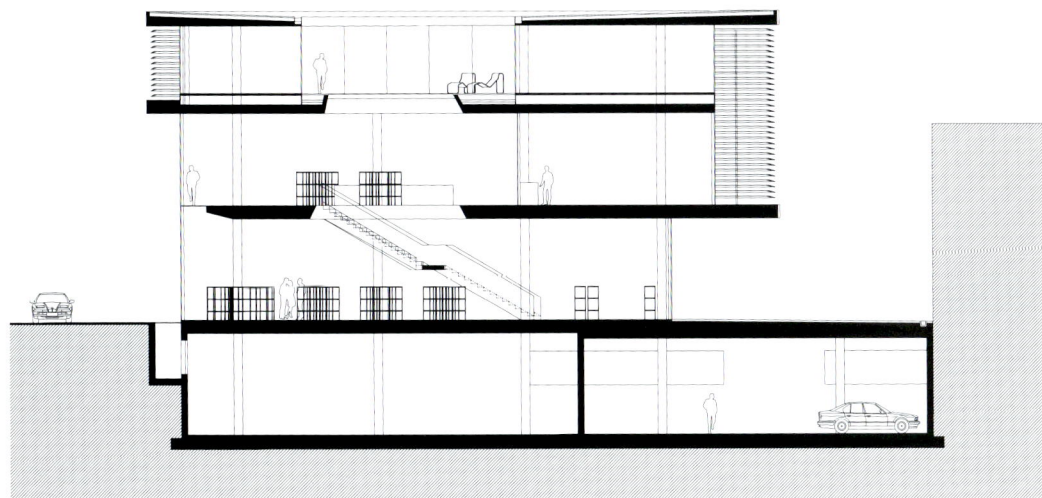

Blaas公司位于意大利Bolzano，是一家以电子机械生产为主的公司。在新的总部办公楼中展示了公司新的产品系列并设有维修服务部门。

大楼的第一层是销售区，第二层是展览区和修理部。所有的管理部门都位于三层。整个建筑结构给人一种单一的封闭之感。公共区域和办公区域之间隔着一个分隔墙，顾客从建筑外面可以清晰地看到这面墙。

北侧玻璃幕墙提供了可视性和透明度，使得展览区和销售区的面积最大化。如维修办公室、仓库和办事处等私人空间的外观都已暴露于南方、东方和西方，但受到了日光隔离系统的保护。

为了建立一个自然光、发展和空间规划的最优关系，设计师在大楼的中心创建了一个明亮的门廊和一个内部庭院。这个绿色的露天场地使得三楼的行政管理部门可以收到充足的自然光，同时为工作人员提供了一个受保护的安静的休闲场所。

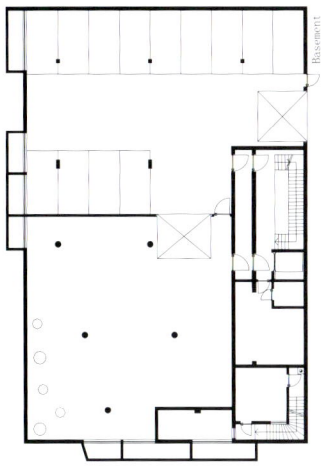

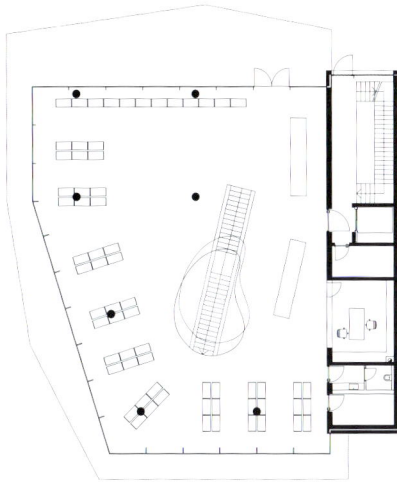

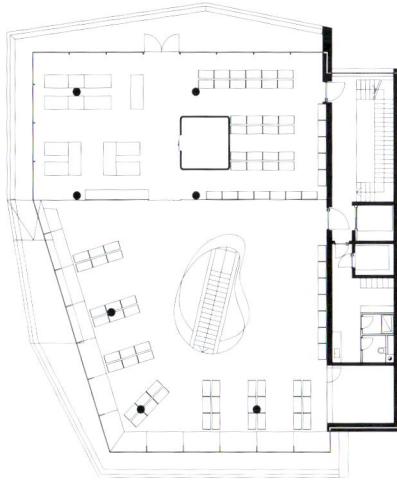

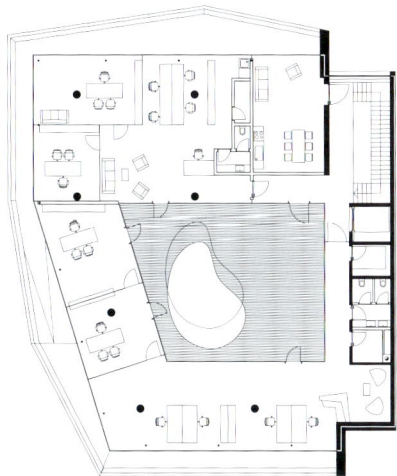

Darwin Center Phase II

Architect: C. F. Møller Architects
Location: Cromwell Road, London
Area: 16000m²
Photographer: Torben

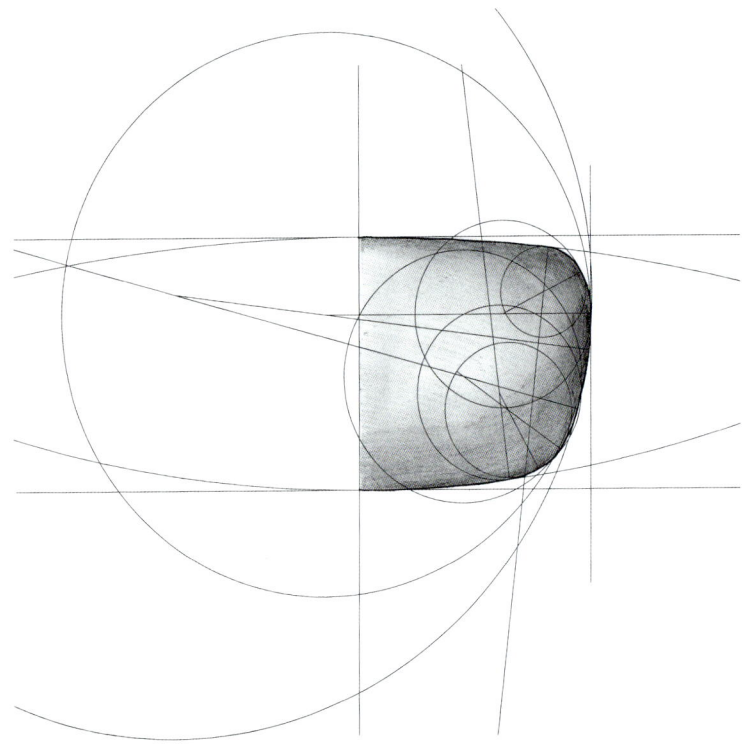

The second phase of the Darwin Center is an extension of the famous Natural History Museum in London, taking the form of a huge eight-storey concrete cocoon, surrounded by a glass atrium. The architecture of the Darwin Center reflects this dual role, and reveals to the public for the first time the incredible range and diversity of the Museum's collections and the cutting-edge scientific research they support.

The centerpiece is made to appear like a large silk cocoon, and forms the inner protective element that houses the museum's unique collection of 17 million insects and 3 million plants. The shape and size give the visitor a tangible understanding of the volume of the collections contained within.

Visitors can experience the Darwin Center as a compelling and interactive learning space, observing the scientific and research activities without interrupting scientific work in progress.

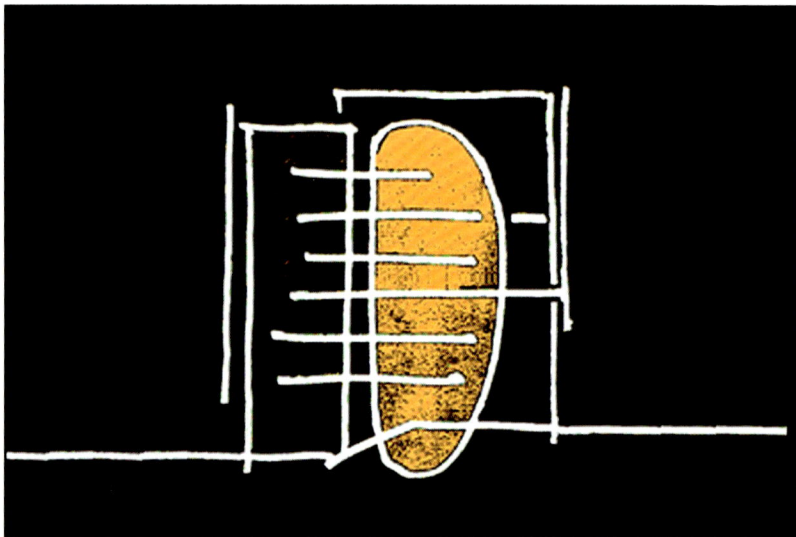

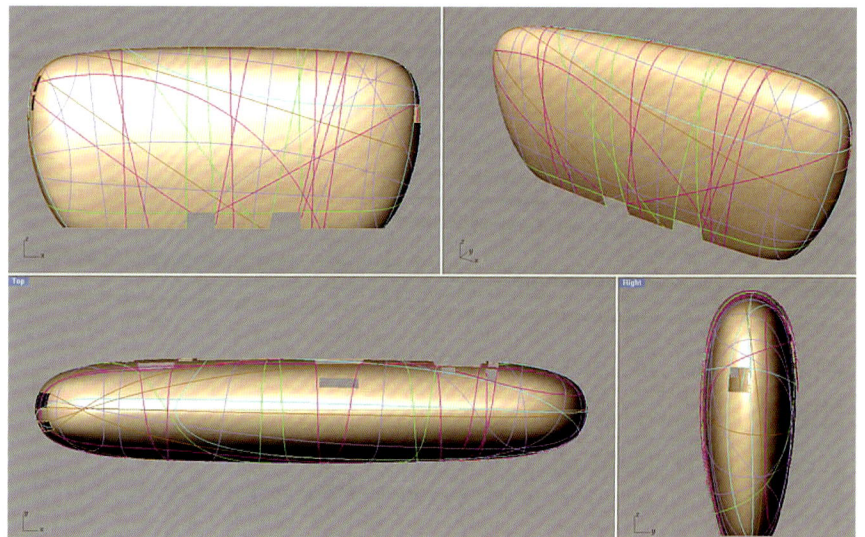

伦敦的国家历史博物馆的扩建是达尔文中心的二期工程。工程采用八层高的混凝土结构包围其外，中间是大玻璃中庭。这个建筑表明了达尔文中心的双重身份，并第一次向大众揭示了品种多样、难以置信的馆内收藏品以及他们资助的科技最前沿的研究成果。

中庭的装饰品看起来就像一个巨大的茧，象征着保护里面珍贵的1700万种昆虫和300万种植物。参观者可以实实在在地感受到里面种类繁多的展览品。

在达尔文中心，参观者可以感受到它是一个让你学习交流的地方。既能观察科学家们怎样进行科学研究，又不会影响到科学研究的进展。

I-way

Architect: Cyrille Druar
Location: Lyon, France
Photographer: Courtesy of Cyrille Druart

Organized around 3 simulation zones, the complex also includes numerous upscale areas for a global experience: fitness room, alcohol-free bar, bar/restaurant lounge with terraces, two conference rooms, meetings rooms and offices.

The building has 3 levels. The main entrance is on the ground floor. The hall leads directly to the second floor via two glass lifts. Access to the first floor is therefore only possible via the upper level. There people can find the main activities of the building.

The building is an architectural work, but also a stage. The site was designed in such a way that something is always happening, every ten seconds or so during a normal visit. The rhythm is engraved in the concrete.

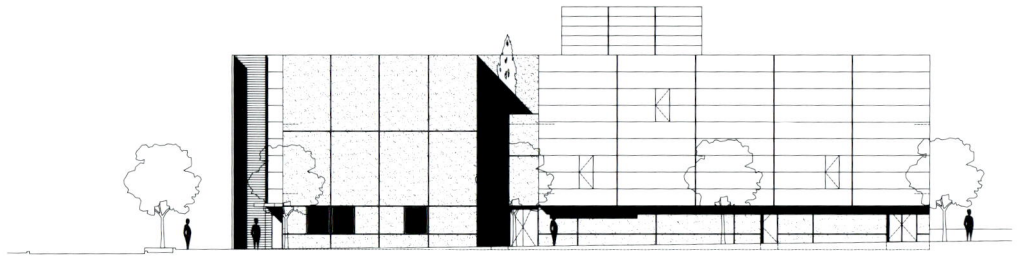

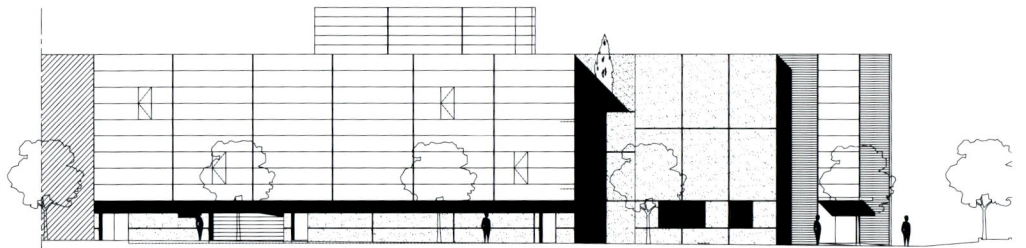

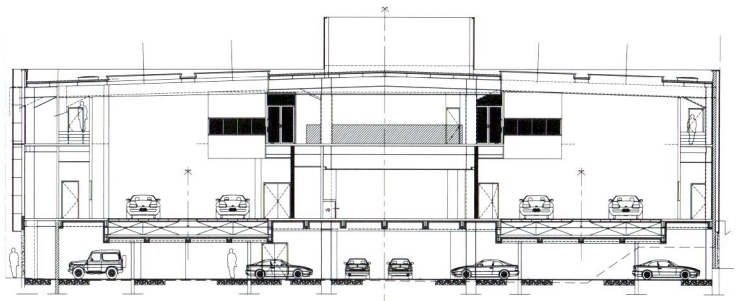

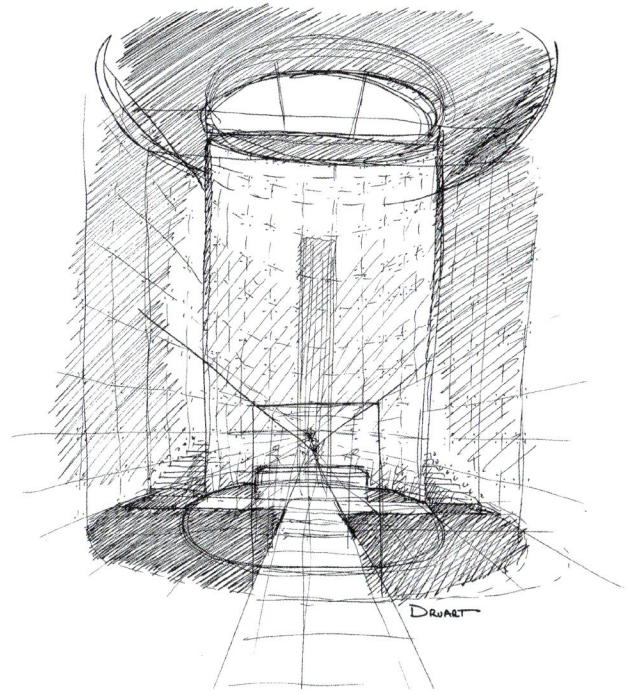

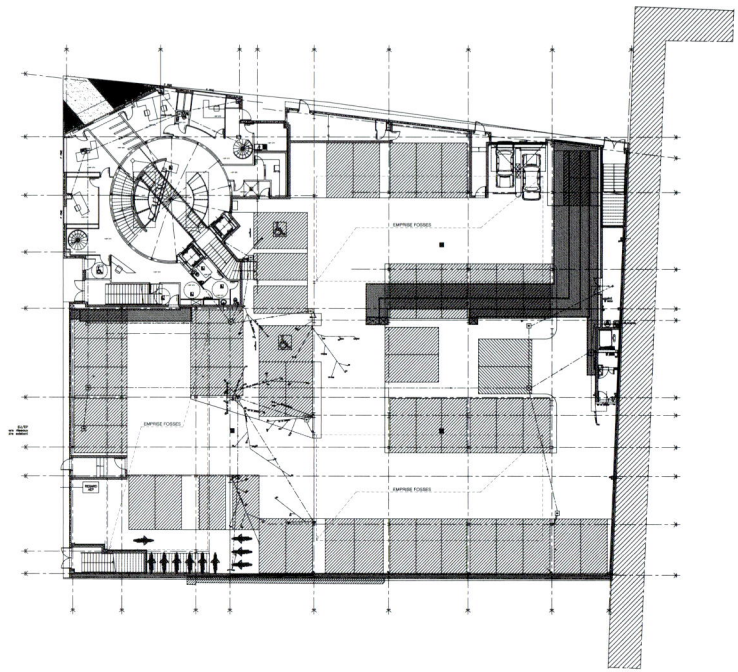

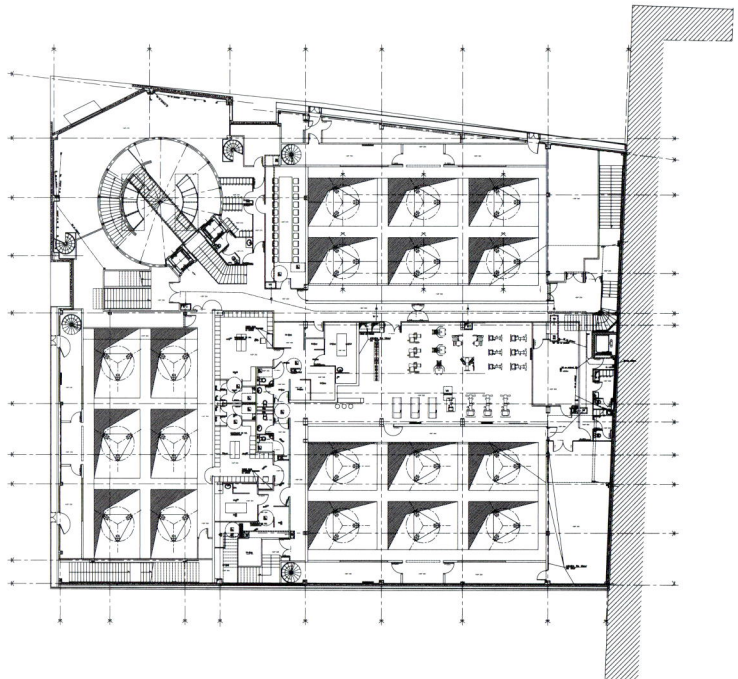

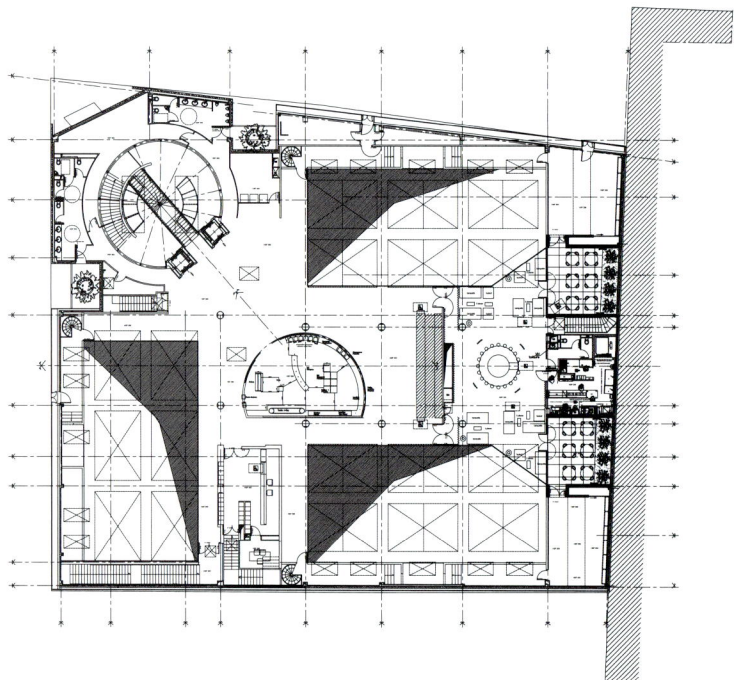

　　这座综合性大楼分为三个区，里面设有各式各样的高档体验场所：健身室、无酒精吧、酒吧、带露台的餐厅酒廊、两间会议室以及办公室。

　　这座大楼共有三层，主入口在一层。穿过一楼的大厅便来到两个透明的玻璃电梯门口，乘坐电梯可以直接到达二层。上面则是该大楼的主要活动场所。

　　这座大楼不仅是一座建筑，更是一个舞台。大楼设计使得它看起来瞬息万变，无论你停留在这座大楼多久，它都能带给你动态感。

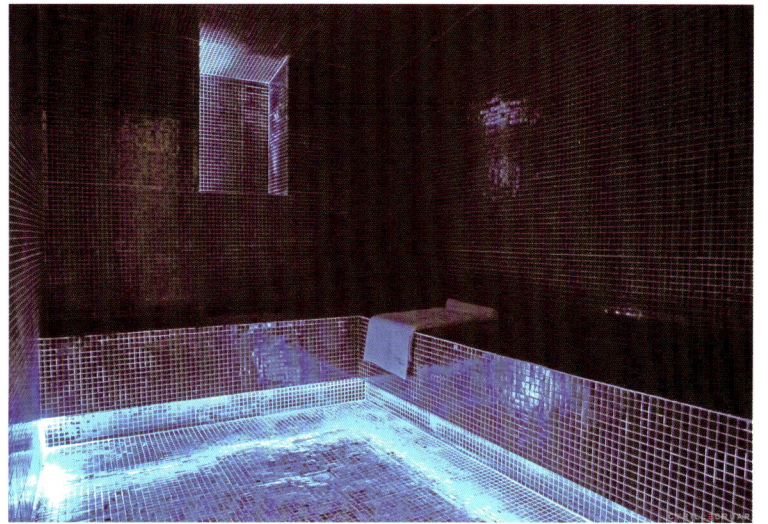

Audenasa

Architect: Antonio Vaíllo i Daniel - Juan L. Irigaray Huarte
Firm: VAÍLLO & IRIGARAY + EGUINOA
Photographer: José Manuel Cutillas

The building offers an image derived morphological genesis of its own: a tablet suspended, almost floating on the gentle slope green slightly twisted repeating the same gestures that the topography offers a gesture of successive concave ribs against the sun: In a flat landscape almost one dimensional, as is the highway, immeasurably longitudinal, from where the building controls and directs the company, contorts, and stands as lookout (also longitudinal), as a new "lookout" observer. Two slabs of concrete lattice steel tape with the corten blocks south and north reused tire. The picture of the complex aims to establish close ties to the movement and infrastructure relating to transport, and perhaps away from the usual urban readings in similar programs.

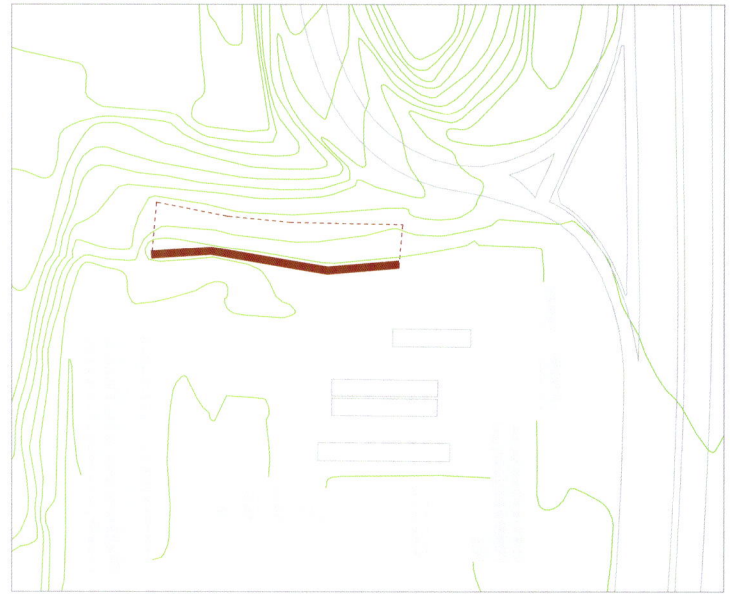

PLANTA SITUACIÓN

N
0 10 20 50m

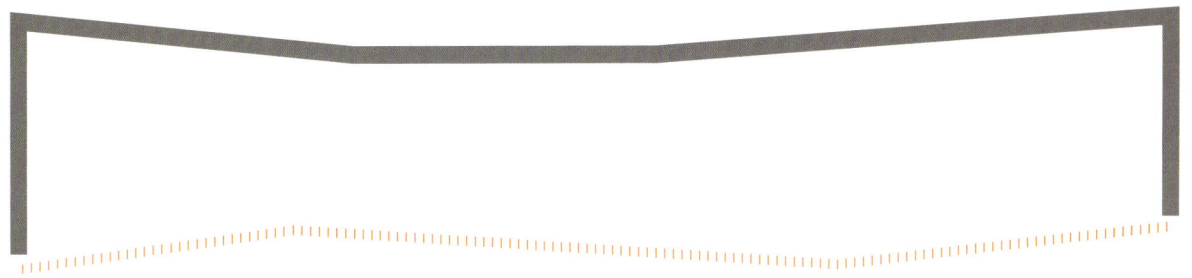

ADAPTACIÓN AL TERRENO

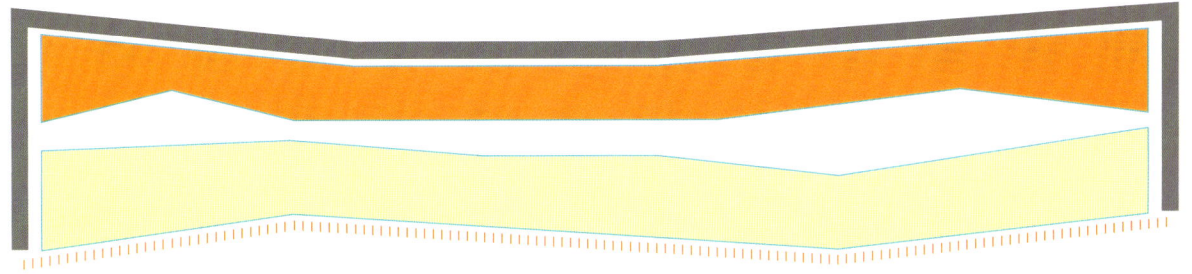

ZONIFICACIÓN USOS

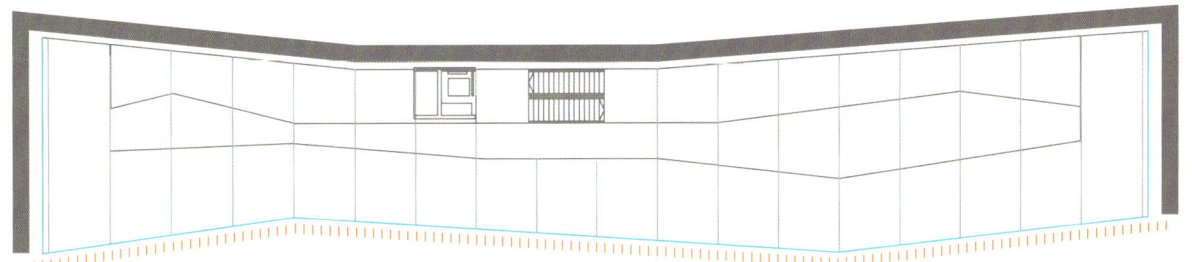

MODULACIÓN

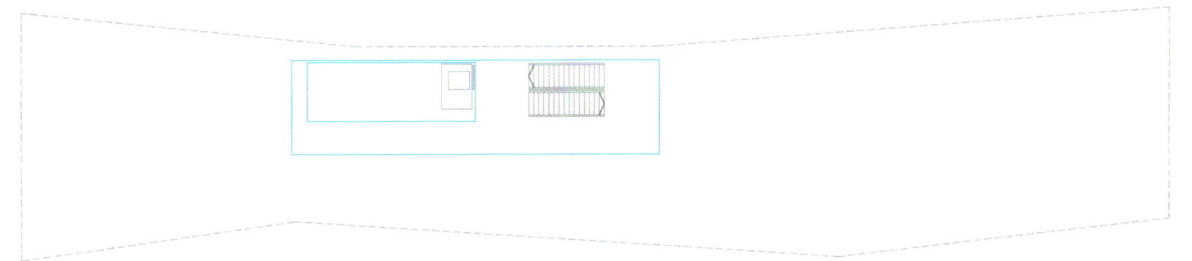

PLANTA

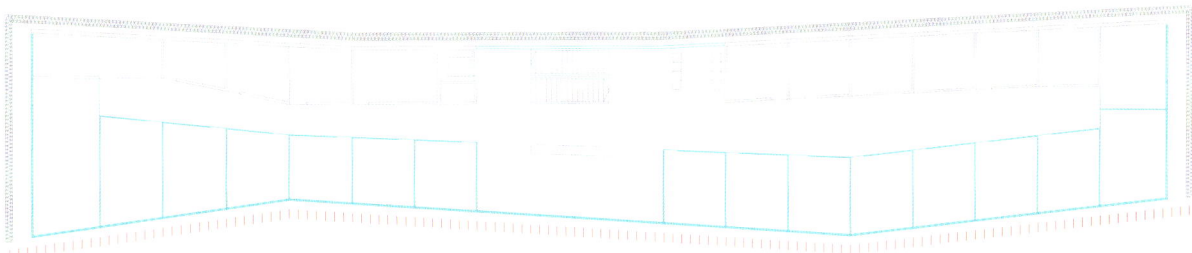

PLANTA PRIMERA e:1/300

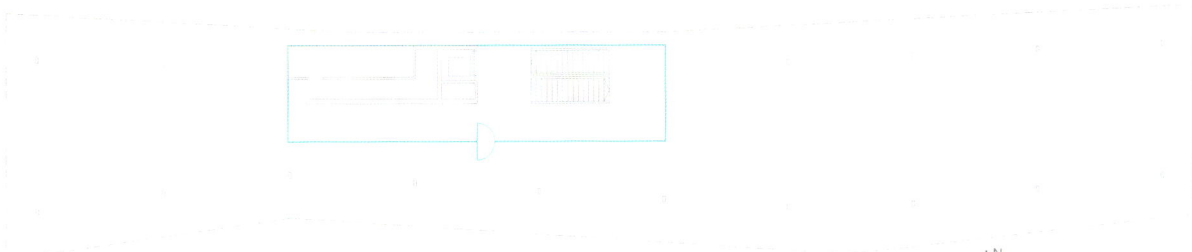

PLANTA BAJA e:1/300

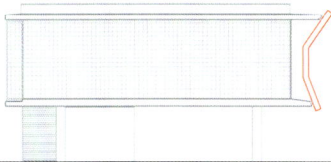

ALZADO OESTE e:1/300

ALZADO ESTE e:1/200

ALZADO NORTE e:1/300

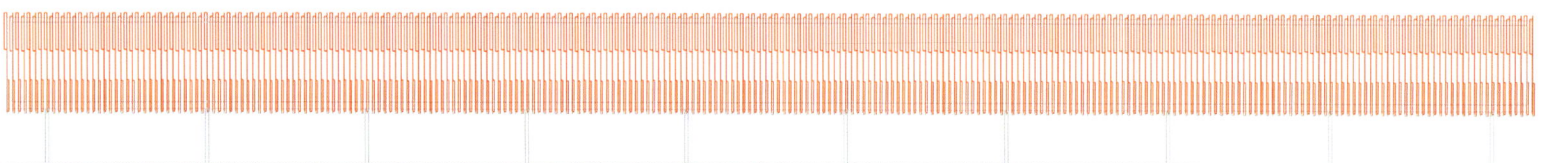

ALZADO SUR e:1/300

0   2   5   10m

DETALLES FACHADA e:1/20

DET-1　　　　　　　　　　　　　　DET-2

SECCIÓN TRANSVERSAL

0　1　2　　　　5m

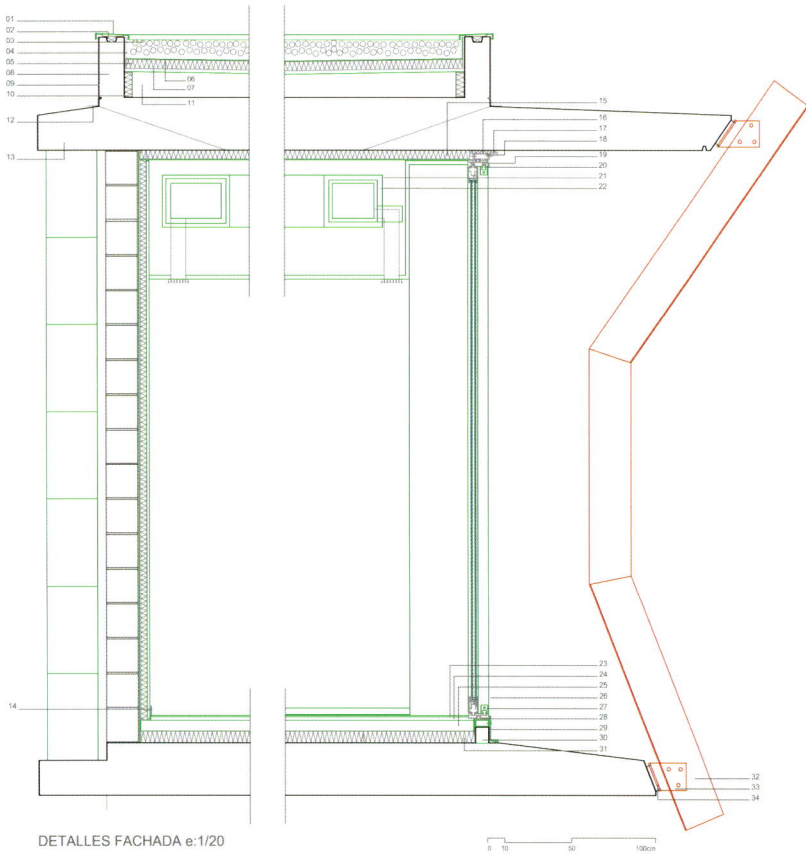

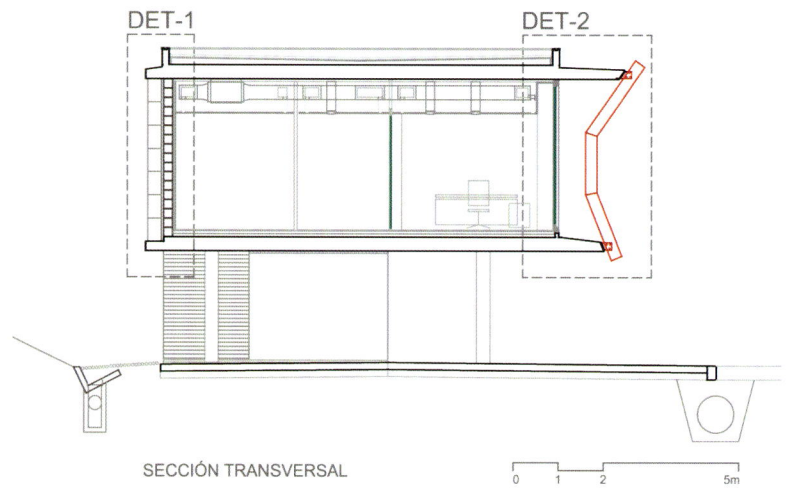

这座建筑提供了源于自身的衍生形态学形象，一张悬浮的写字板几乎飘浮在绿色的缓坡上，以同一姿态的轻微重复扭曲，成功地顺应地势形成凹面肋骨的姿态来抵挡太阳光。在这个几乎一维的景观上，一栋处在广阔无边的高速公路旁的建筑，控制和引导着它的同伴，并由纵向的瞭望台来作为新的观察模式。两块网格状混凝土钢板与南面的耐腐蚀板和北面用循环再用的轮胎装饰形成的面板连接一体，这个综合设施的设计蓝本在行动和基础设施之间建立起与运输有关的亲密的联系，也许能在类似的项目中建立起不同寻常的建筑形式。

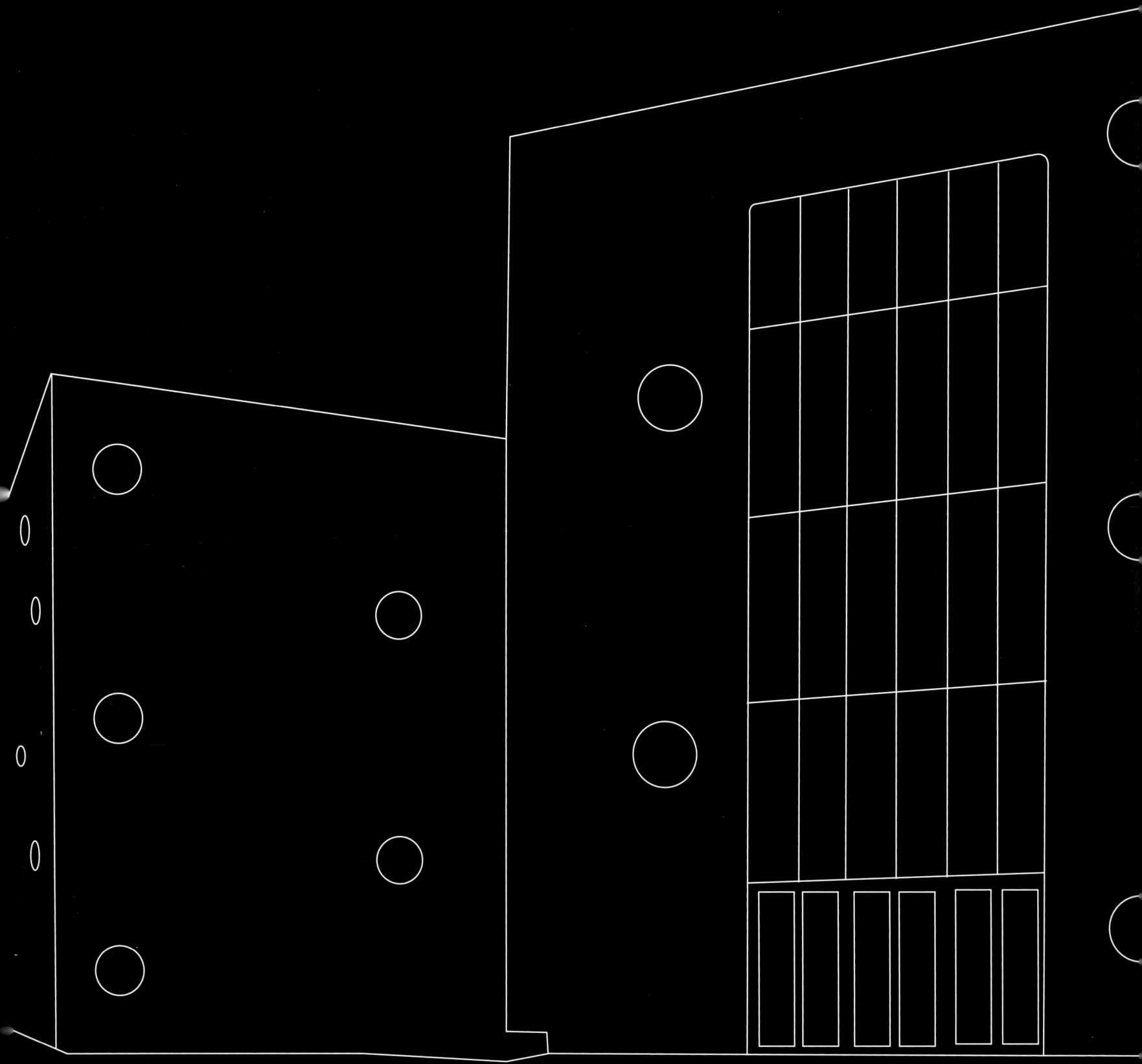

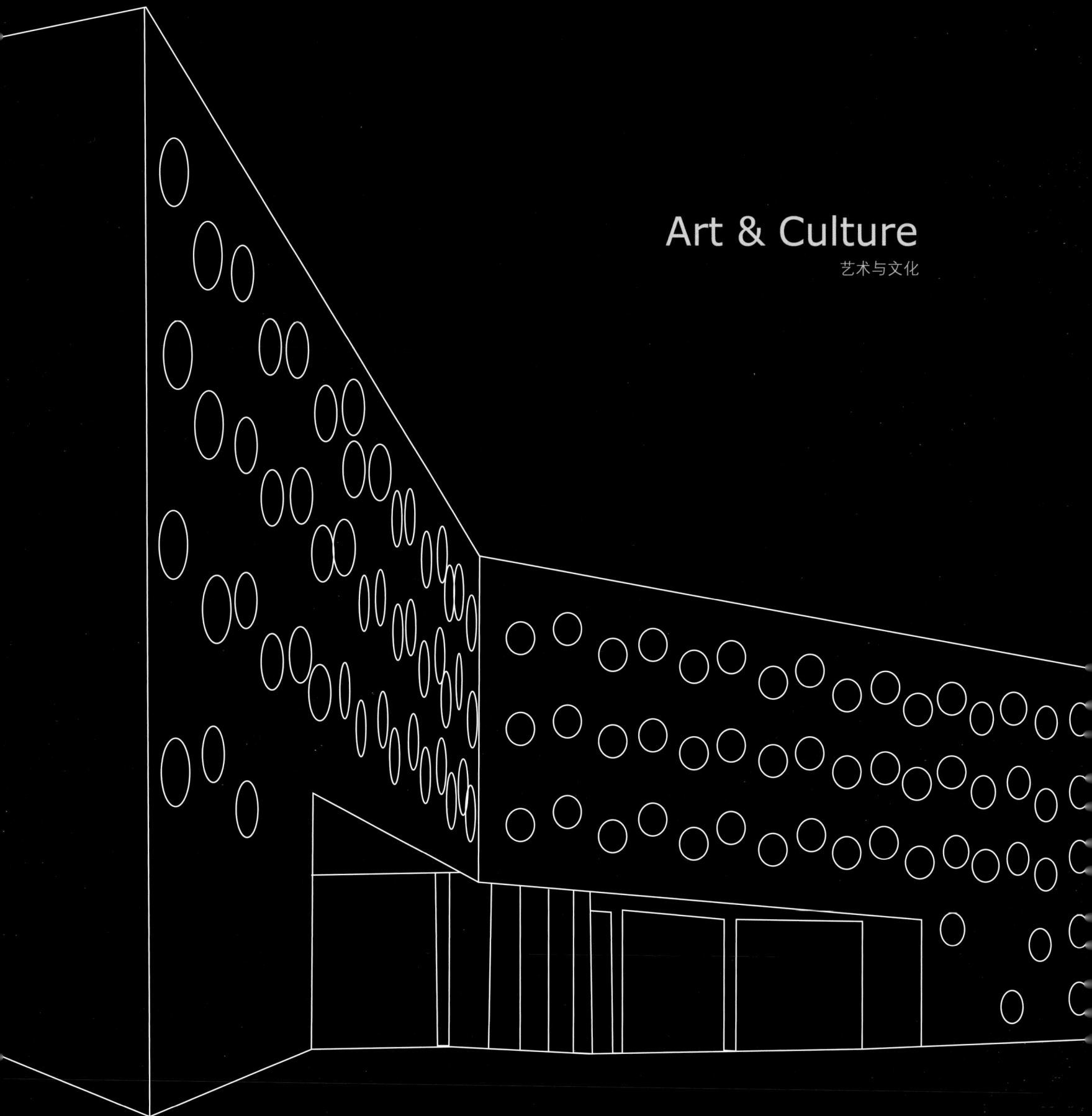

Art & Culture
艺术与文化

The Extension to the Denver Art Museum

Architect: Daniel Libeskind
Firm: Studio Daniel Libeskind
Landscape Architect: Studio Daniel Libeskind with Davis Partnership
Location: Denver, Colorado, USA
Area: 16723m²
Photographer: Studio Daniel Libeskind, Bitter Bredt Fotografie, Michele Nastasi,
Denver Art Museum

The Extension to the Denver Art Museum, is an expansion and addition to the existing museum, designed by the Italian architect Gio Ponti.
The Hamilton Building's design recalls the peaks of the Rocky Mountains and geometric rock crystals found in the foothills near Denver. The materials of the building closely relate to the existing context as well as introducing innovative new materials, such as the 9,000 titanium panels which cover the building's surface and refect the brilliant Colorado sunlight.

The new building is a kind of city hub, tying together downtown, the Civic Center, and forming a strong connection to the golden triangle neighborhood. The project is not designed as a stand-alone building, but as part of a composition of public spaces, monuments and gateways in this developing part of the city, contributing to the synergy among both large and intimate neighboring spaces.

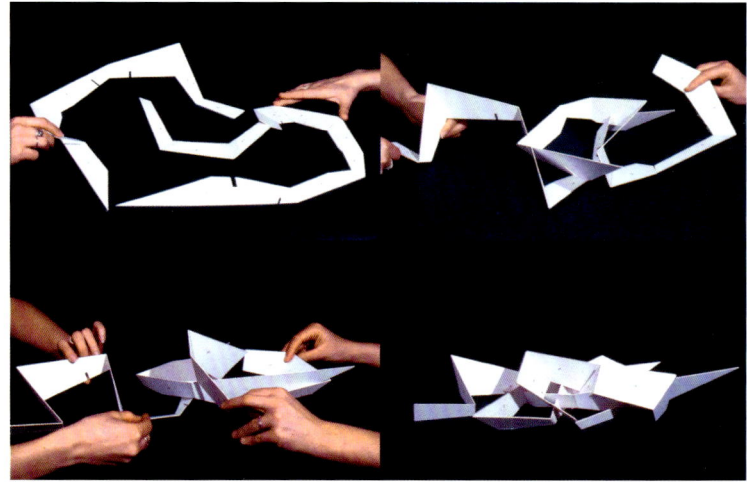

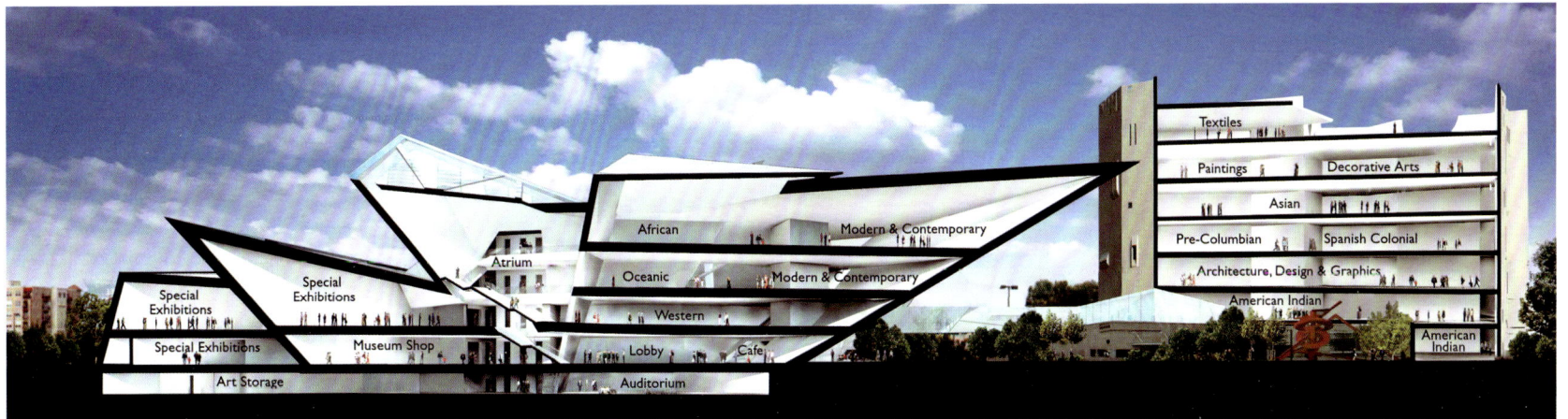

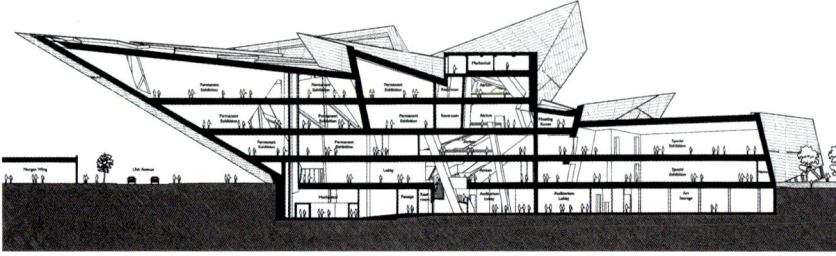

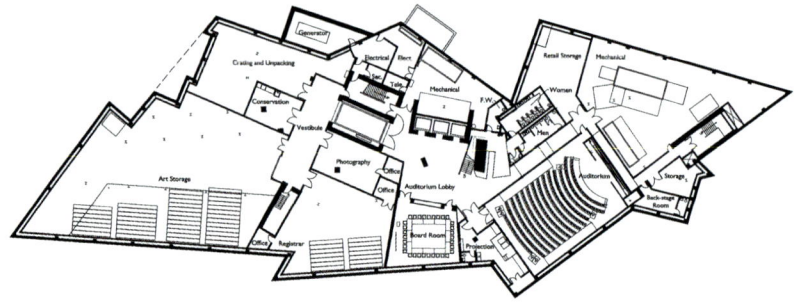

Basement level floor plan

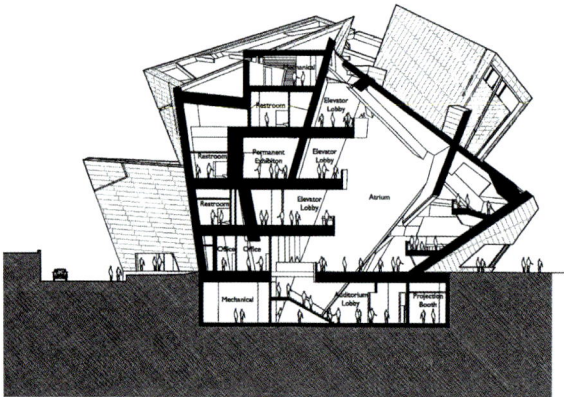

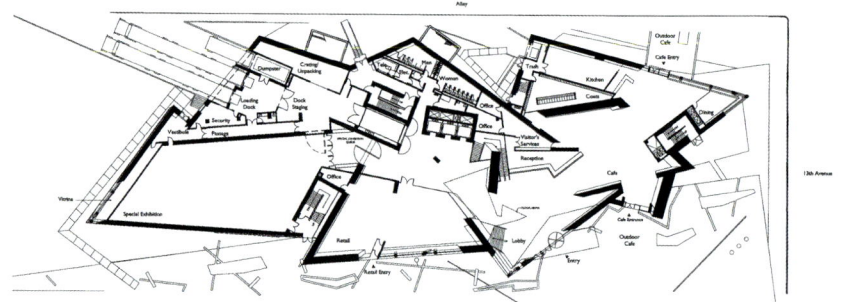

First floor plan

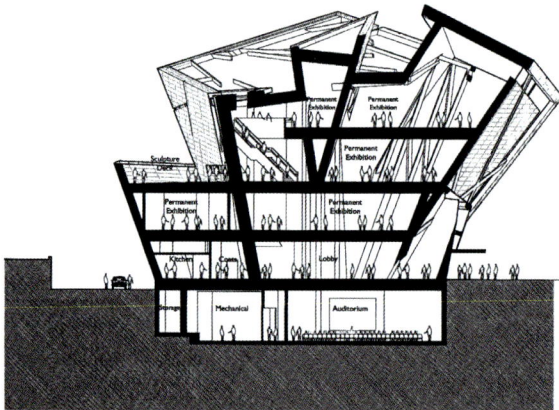

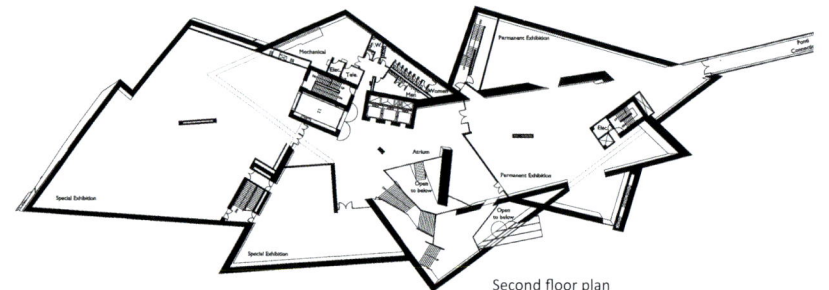

Second floor plan

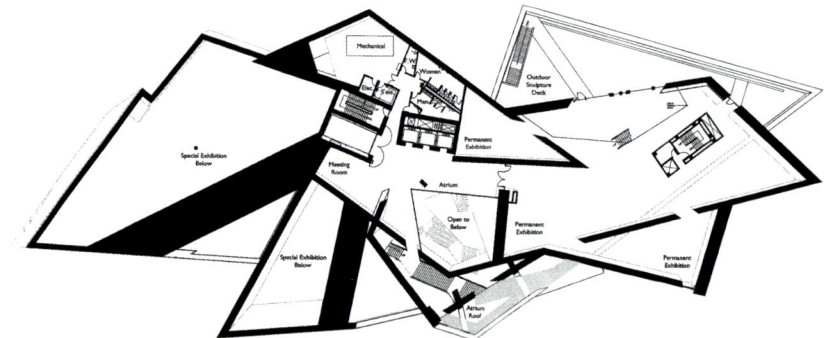

Third floor plan

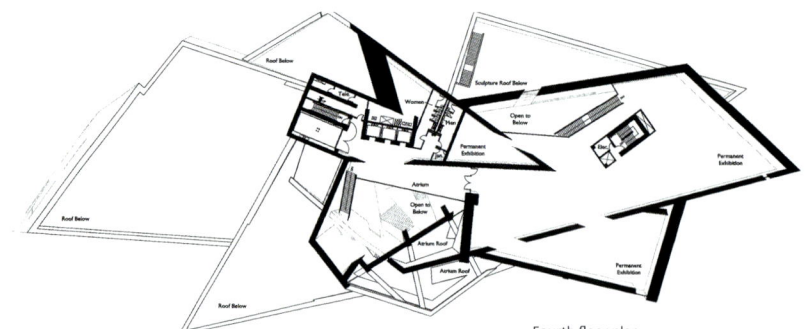

Fourth floor plan

丹佛艺术博物馆的扩建工程是对现存博物馆的扩大与增建，由意大利建筑师吉奥·庞蒂设计。

汉密尔顿建筑物的设计让人想起了落基山脉的山峰和在丹佛市附近发现的几何岩石晶体。建筑材料与背景环境密切相关，同时引入新的材料，如覆盖整栋建筑物表面并反射科罗拉多州灿烂的阳光的9000块铁板。

这座新楼位于城市中心，与商业中心和市政中心连接在一起，形成邻近区域中的一个金三角。这个项目并不是一个独立的建筑，因其与周围广阔且亲密的邻近空间的共同作用，使这栋建筑成为这个发展中的城市的公共空间、不朽的作品和城市网关的组成部分。

Surry Hills Library and Community

Architect: Richard Francis-Jones
Firm: Francis-Jones Morehen Thorp
Location: Surry Hills, New South Wales, Australia
Photographer: John Gollings, Andrew Chung, Mathieu Faliu

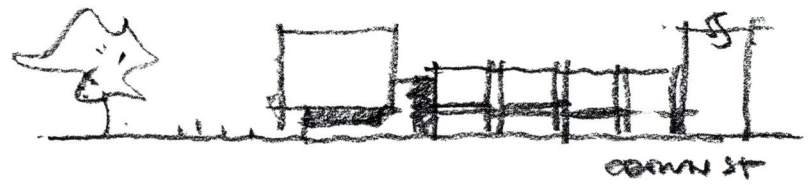

This project is prominently located in the heart of Surry Hills, an inner-city suburb of Sydney whose community is characterized by a diversity of age, income and cultural backgrounds. The architectural context is also diverse: residential apartments, terrace housing, shops and commercial/industrial premises, vary in scale though their architectural style is predominantly Victorian. The site is very constrained, measuring just 25 by 28 metres and bound on three edges by roads: Crown Street, the main street of Surry Hills, to the east and two residential streets to the south and west.

The project's brief was developed in close consultation with the very active local community. The key approach was that the community wanted a facility that everyone could share. Rather than only a library or a community center or childcare center, it became clear that it was important to have all of these facilities together in one building, in one place. In this way the building became a truly shared place where the whole community could meet and use in different ways. Important, too, was for the building to represent and reflect the community's values.

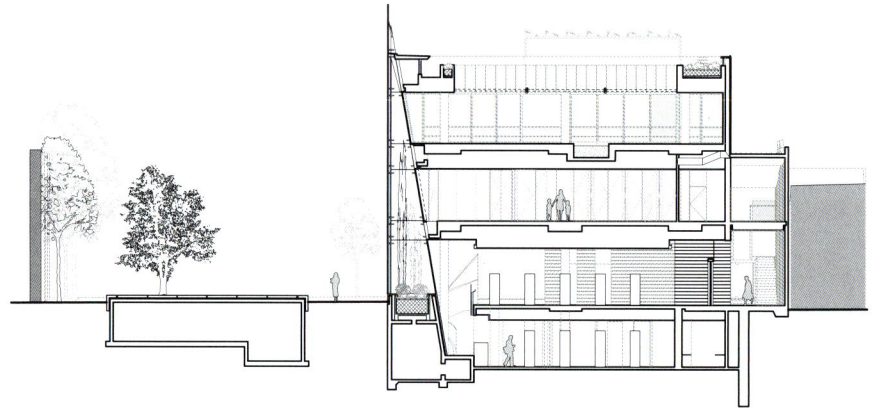

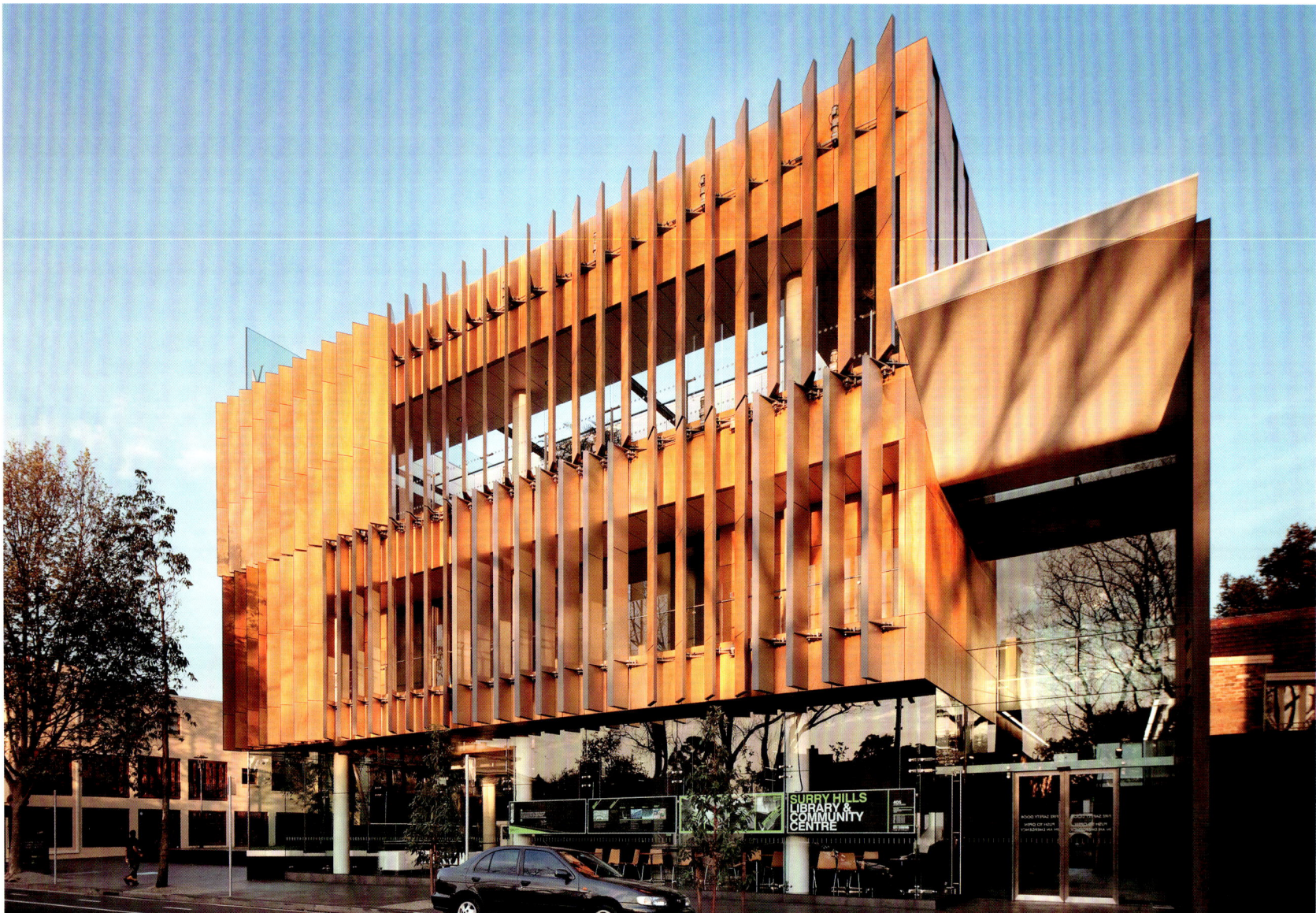

该项目位于悉尼市郊的萨利希尔斯，非常引人注目。这个社区以年龄、收入、文化背景的多样性为特色。建筑类型也同样多样化：住宅式公寓、联排式住宅、店铺、商业及工业房屋，规模各异，但都是维多利亚风格（夸张但适宜）。该建筑场域很有限，25米长，28米宽，由公路通向三个区域：皇冠大道，萨利希尔斯大街，到达东面及南面和西面的住宅区街道。

项目设计前经过与当地社区居民的商量，最后讨论得出的一个关键结论是：居民们想要一个每个人都能享用的设施，不是仅有一个图书馆、社区中心或托管中心。很明显，居民们所希望得到的是在同一个地方的一栋建筑物里集中所有这些设施。这样，这栋建筑物不但能成为整个社区居民可以聚会并使用的一个共享的地方，也能代表并反映社区的价值观。

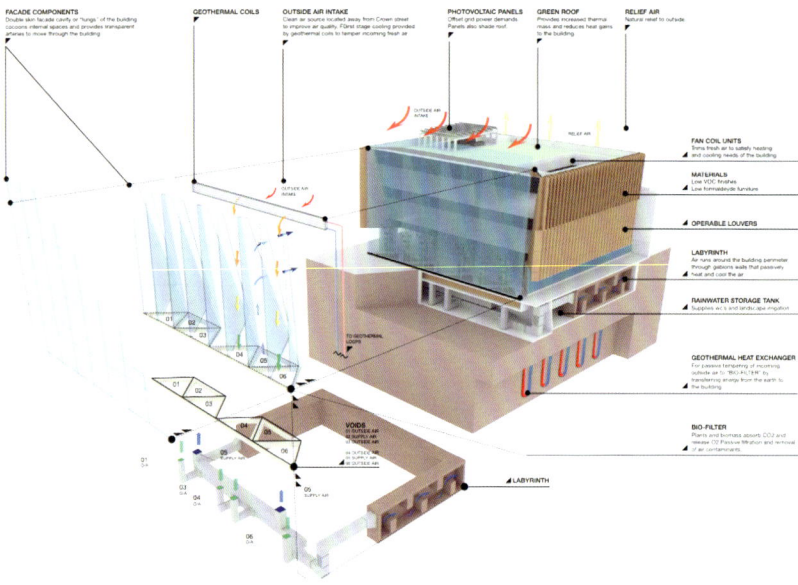

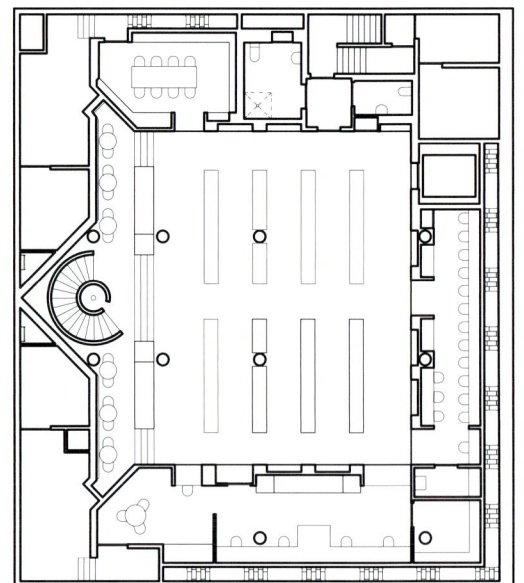

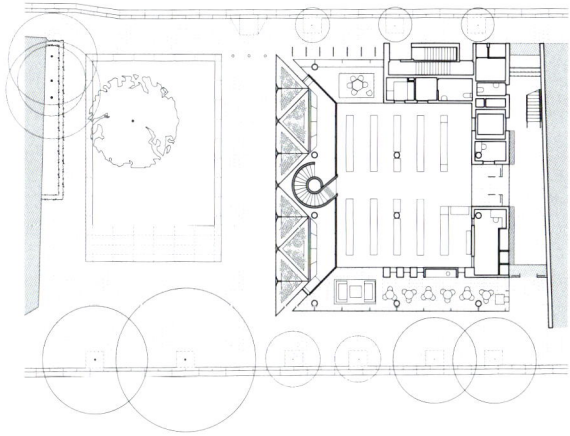

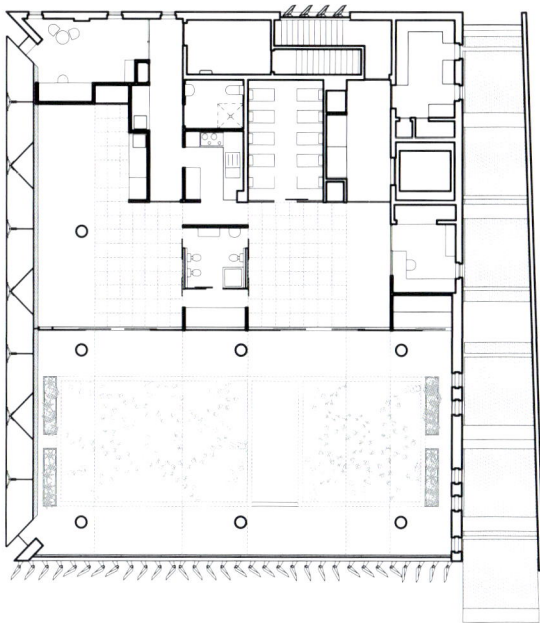

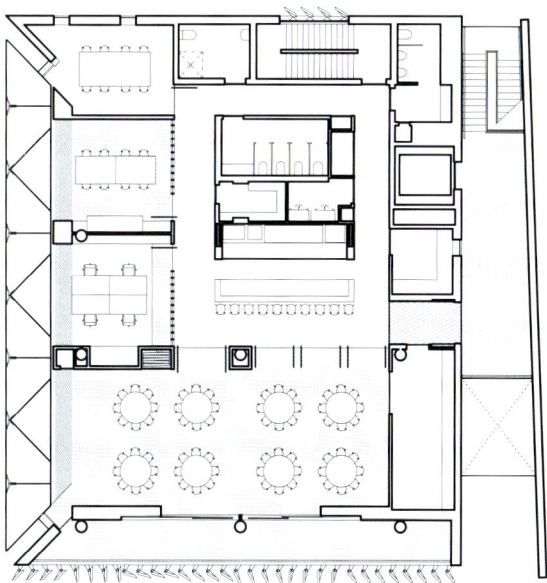

Bijlmer Park Theater

Architect: Paul de Ruiter
Team: Willem Jan Landman, Dieter Blok
Location: Anton de Komplein, Amsterdam
Area: 1953m²
Photographer: Pieter Kers

The cultural building consists of an ellipse shape, with the upper two floors slightly displaced in relation to the ground floor. The elliptical shape of the building did mean that it was necessary to search for a financially viable way of reproducing t his rounded shape in the partially glass façade. The solution was found in a combination of wooden slats and vertical aluminium strips placed against the steel and glass sections of the facade. This means that the intersection points of the segmented façade are not visible and the building has a rounded, dynamic and somewhat abstract appearance that changes continually as you walk around it.
During the day, the striking shape of the cultural building makes it clearly recognizable, while it is conspicuous in the evening because of its color, which can be altered to fit the occasion. This is made possible by the use of LED lighting. A line of light is fitted behind the steal façade in the façade, shining downwards. Because this light shines against the steel façade and the wooden slats, the building acquires an appearance of transparency, as if the light is coming from inside the building.

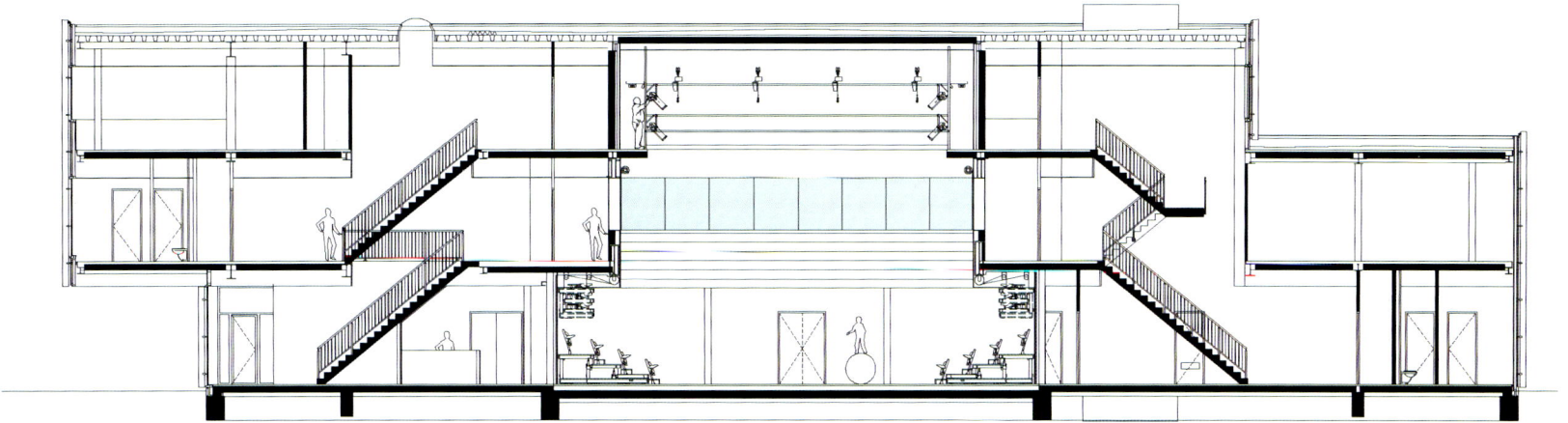

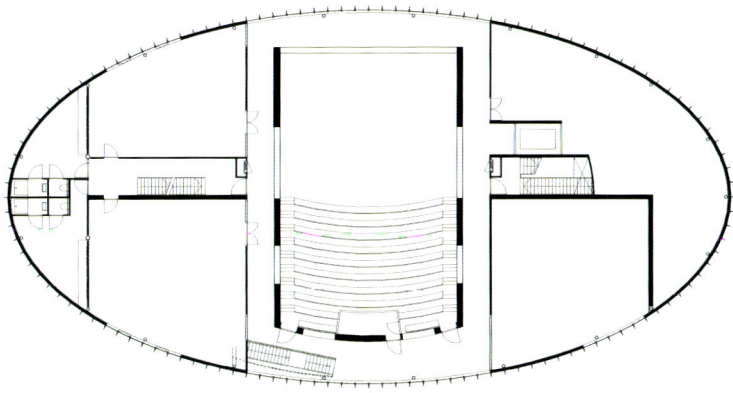

scale 1:200
first floor

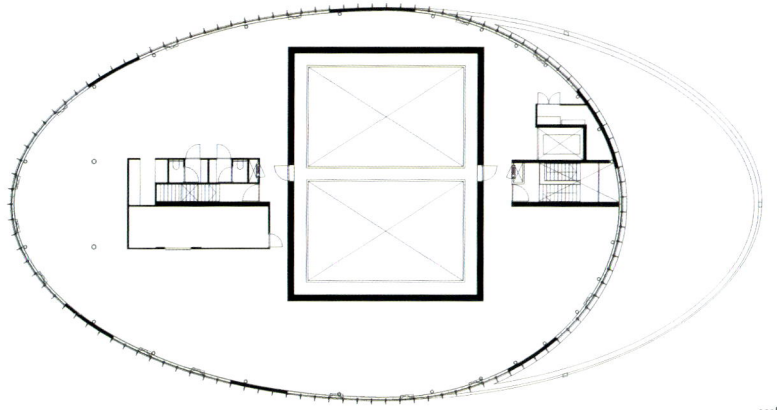

scale 1:200
second floor

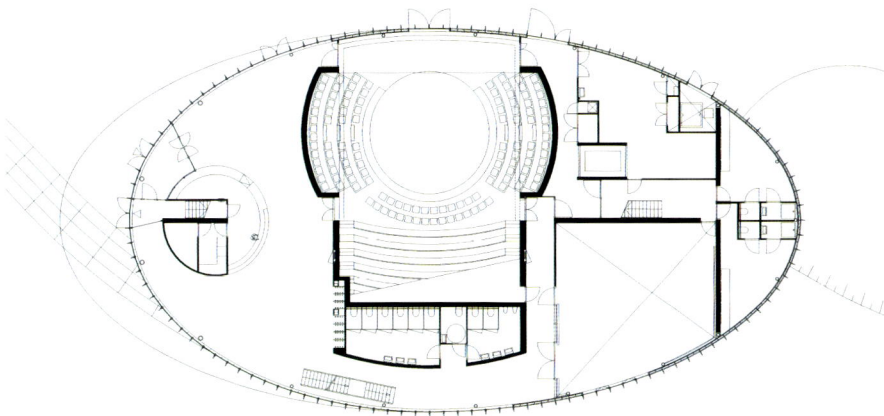

scale 1:200
ground floor

该建筑由一个椭圆形组成，上面的两层和底层有点错位。椭圆形的体型意味着需要找到一个经济可行的方法来解决部分玻璃立面的弧形问题。解决方法是：运用木条和竖向铝片构件覆盖在钢和玻璃组合的立面上面。这样做的效果是分段的立面交接点不见了，这样建筑看上去圆滑、动感，有着抽象化的统一形象，并且当你围绕它走动时，每个角度给人的视觉感受都不同。

白天，这个剧场显著的体型使得它清晰、易于辨认；而晚上它因为自身的色彩同样引人注意，并且这些色彩可以根据不同的活动而变化。在金属立面的后面和里面安装有成条的由发光二级管制成的灯，能依次点亮。正因为金属立面和木条后面闪亮的灯光，使整个建筑获得了一个半透明的效果，就好象灯光是从建筑物内部射出来的一样。

Tampa Museum of Art

Architect: Stanley Saitowitz
Firm: Stanley Saitowitz / Natoma Architects
Location: Tampa, Florida
Area: 6132m²
Photographer: Richard Barnes, James Ostrand

A glass pedestal supports the jewel box of art above. The building floats in the park, embracing it with its overhanging shelter and reflective walls. The lobby is at first horizontal, with entirely glass walls, two sides clear, two sides etched. The clear walls allow the site to run through the space, linking the Performing Art Building on the north with the turrets and domes of the University of Tampa on the south. Above the glass, the perforated ceiling wraps from the exterior into vertical perforated walls that turn into an upper ceiling, perforated again by a series of skylights. Through the dramatic image stairs in the lobby you can arrive the picture gallery in the second floor.

The designer has built the most expansive and generous field of galleries as instruments to enable, through curation, a world to expose art. They are arranged in a circuit, surrounding the vertical courtyard void. The galleries are blank, walls, floor and ceiling all shades of white, silent like the unifying presence of snow. The floors are ground white concrete with a saw cut grid to echo the illuminated white fabric ceiling above. Linear gaps in the ceiling conceal sprinklers, air distribution and lighting.

The outer walls of the bulding is one of the characters of the museum. By day the surfaces reflect the light, with moray like images of clouds or water or vegetation. By night the exterior becomes a canvass for a show of light. It is the magical illumination of the skin changing colors and patterns in endless variations. Take the dazzling exhibits outdoor and display to the whole city then the infinite light and shadow are invested into the river.

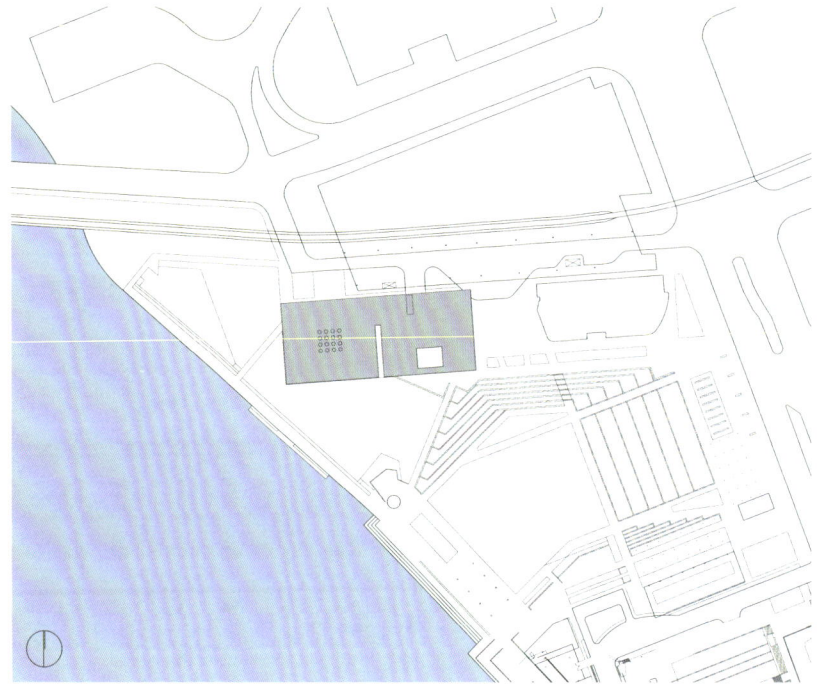

SECTION C-C   0'   10'   50'

SECTION D-D   0'   10'   50'

SECTION A-A   0'   10'   50'

SECTION B-B   0'   10'   50'

8" DUCT W/
INSULATION AND
LINER PANEL.
PAINT WHITE -
SEE MECHANICAL

LIGHTING TRACK; AMBIENT
LIGHT AND SPOT LIGHT

FIXTURE

LUMINOUS CEILING FRAME
STRAPS FOR BRACING; PAINT
WHITE AND ANCHOR TO PLANK

LUMINOUS CEILING
-2 LAYERS (BOTTOM
TRANSLUCENT / TOP CLEAR)

U.S. CEILING 41'-4"
LEVEL 2

SPRINKLER; HEAD AND LINE TO BE WHITE.
ADDITIONAL HEAT COLLECTOR: 14"x12", 18 GA.
SHEET-METAL. PROVIDE WHITE ESCUTCHEON AT
PIPE PENETRATION. PAINT WHITE.

10'

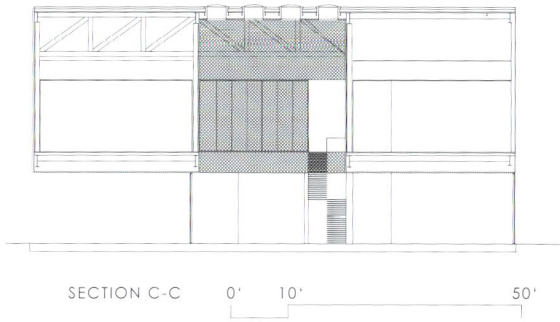

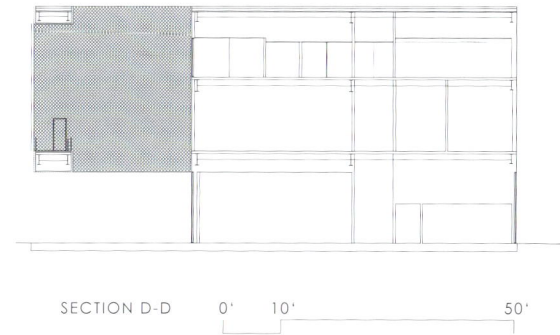

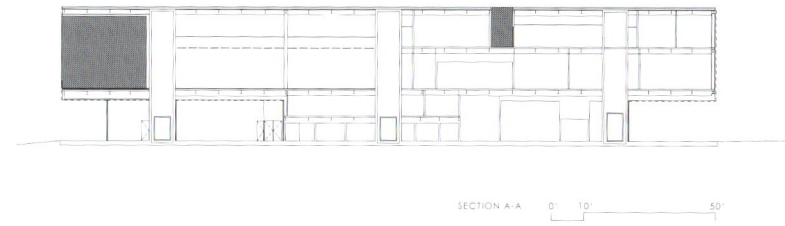

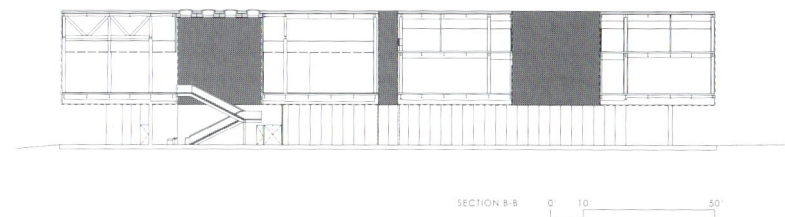

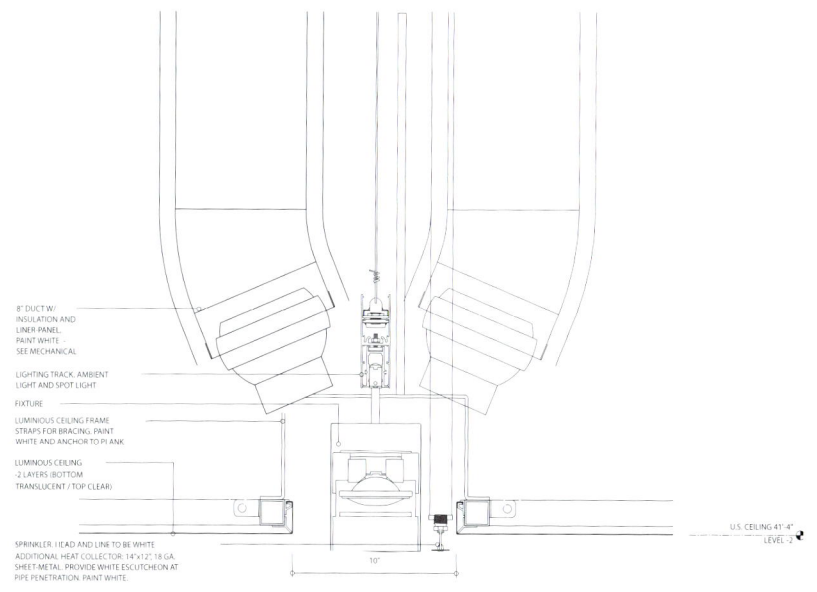

TYPICAL GALLERY LIGHTING STRIP

103

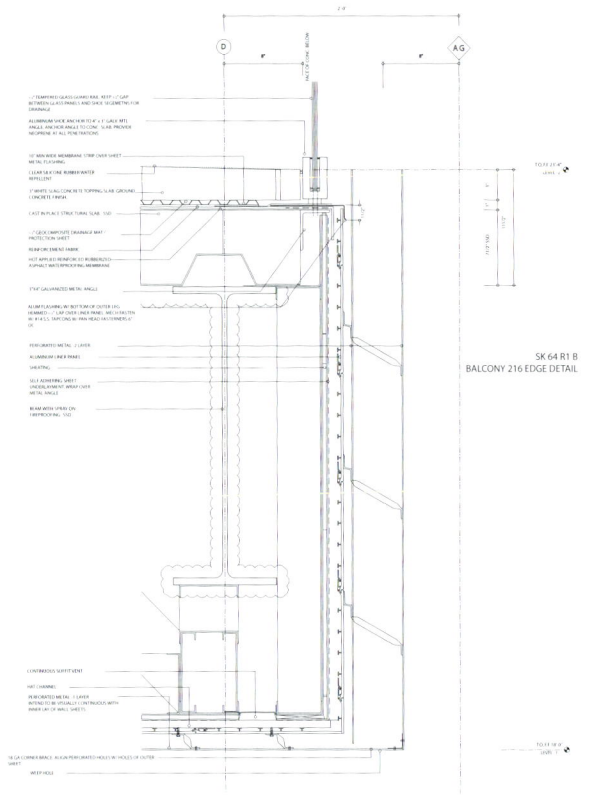

SK 64 R1 B
BALCONY 216 EDGE DETAIL

BRIDGE EDGE

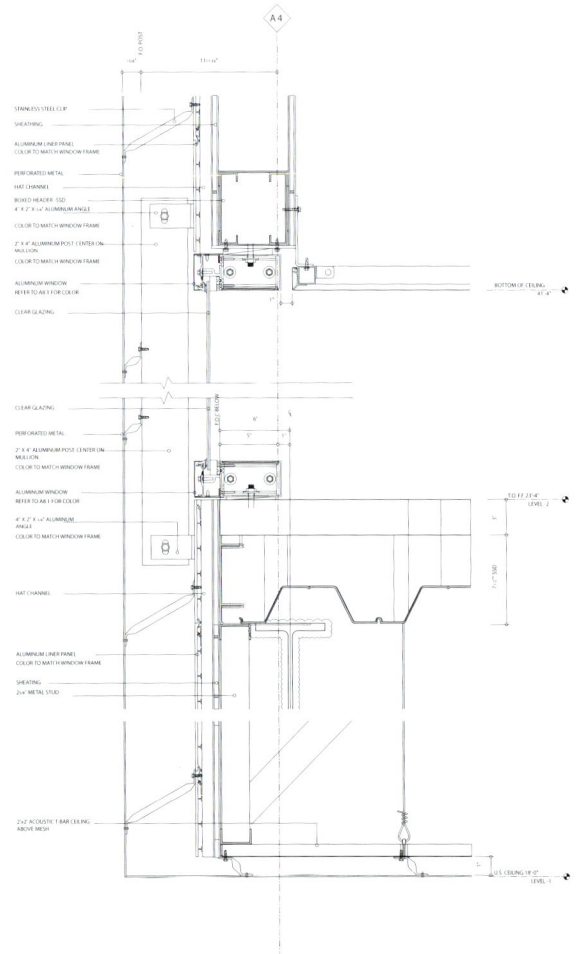

EDGE DETAIL AT ATRIUM SPACE

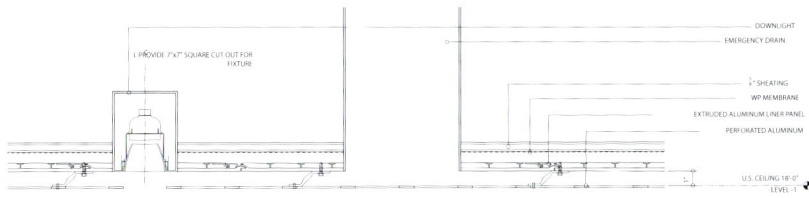

TYPICAL EXTERIOR SOFFIT

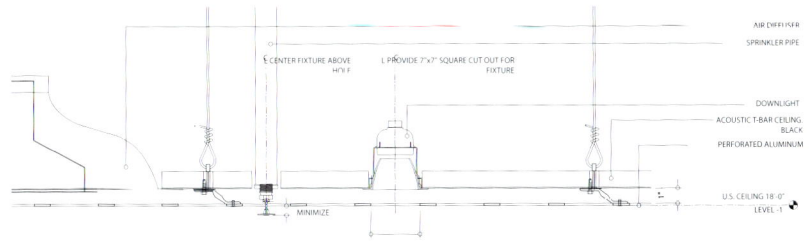

TYPICAL INTERIOR SOFFIT

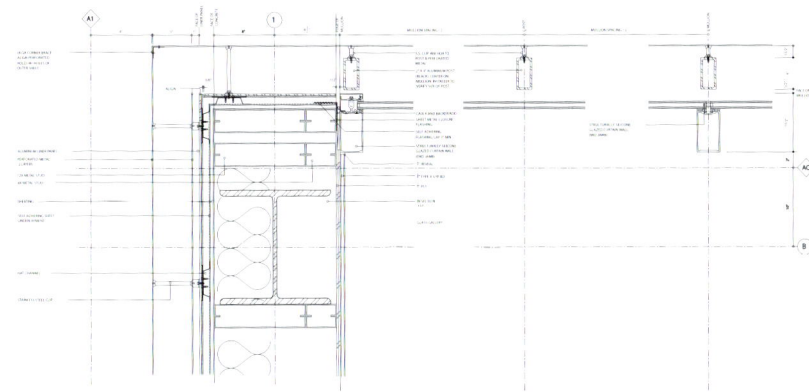

CURTAIN WALL CORNER

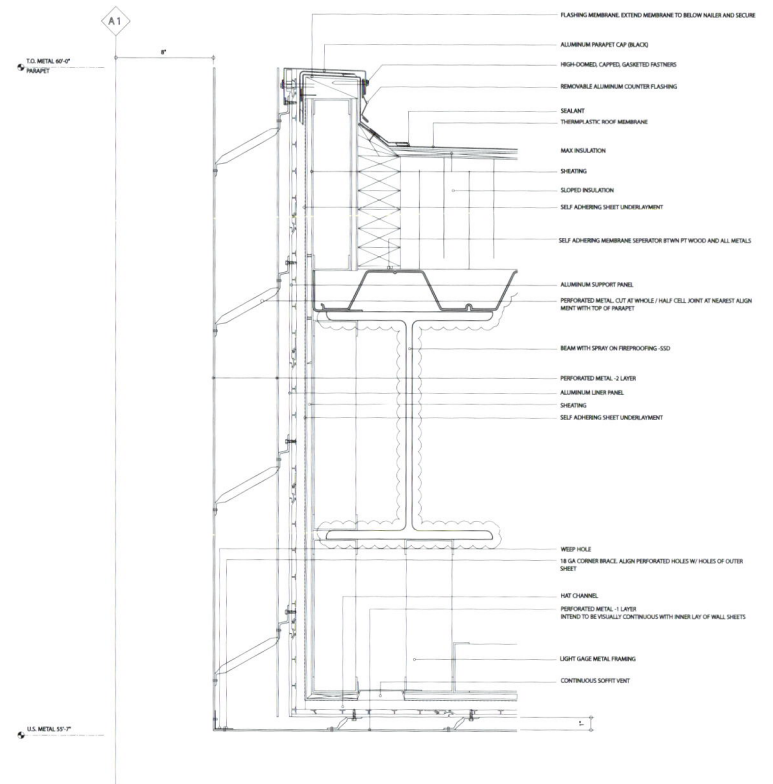

PARAPET DETAIL AT BALCONY

坦帕艺术博物馆犹如一个玻璃基座，承托着上方的艺术"珠宝盒"，被可反光的外围护包围着的博物馆如同飘浮在公园里。一层大厅全部由玻璃墙围合而成，两面透明，两面磨砂。透明玻璃墙将外部景观引入室内，透过玻璃墙，可以看见北面表演艺术中心的塔楼以及南面坦帕大学的圆顶。玻璃之上的部分全由穿孔铝板包裹，通过大厅里的极富戏剧色彩的影像楼梯可达二层的画廊。

设计师慷慨地把最好的场地用于打造画廊空间，让其成为一个展示艺术品的世界。画廊环绕中庭，墙壁、地板和顶棚均涂以白色，给人一种雪地无声的感觉。地面是切割出网格的白色混凝土，以呼应白色的顶棚。顶棚上线形结构的沟槽隐藏了灭火喷头、换气孔和照明设备。

该博物馆的外墙也是其特色之一，白天，光折射在表面上，云、水或植物在其上流淌，仿佛海市蜃楼。夜晚来临时，人们则会看到一条不断变幻的彩色缎带，整个建筑不断变换着颜色和形态，把室内的琳琅满目呈现到室外，展示给整个城市，又把无穷的光影投入河水中。

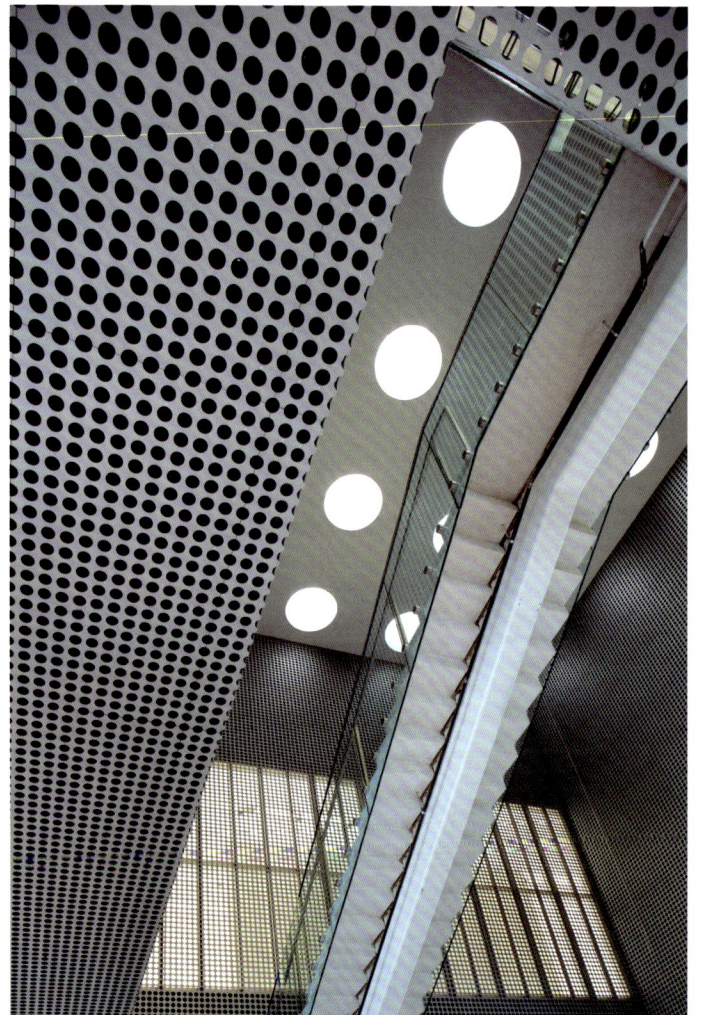

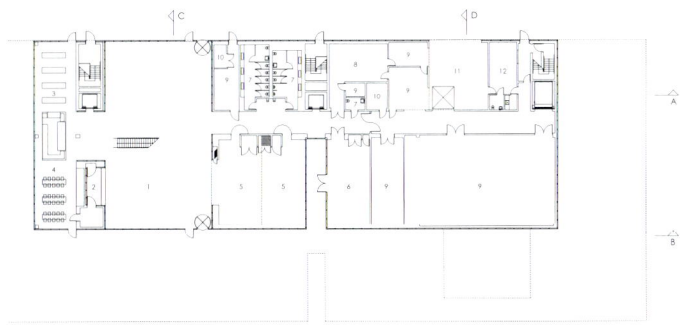

GROUND FLOOR   0' 10'   50'

1. LOBBY AND RECEPTION	12. SECURITY
2. TICKET DESK	13. GALLERY
3. MUSEUM STORE	14. BALCONY
4. MUSEUM CAFE	15. ATRIUM
5. CONFERENCE ROOM	16. WORKSHOP
6. CLASSROOM	17. STAGING
7. RESTROOM	18. REGISTRAR
8. CATERING KITCHEN	19. RECEPTION
9. STORAGE	20. OFFICE
10. MAINTENANCE	21. KITCHEN
11. LOADING DOCK	22. BOARDROOM

SECOND FLOOR   0' 10'   50'

1. LOBBY AND RECEPTION	12. SECURITY
2. TICKET DESK	13. GALLERY
3. MUSEUM STORE	14. BALCONY
4. MUSEUM CAFE	15. ATRIUM
5. CONFERENCE ROOM	16. WORKSHOP
6. CLASSROOM	17. STAGING
7. RESTROOM	18. REGISTRAR
8. CATERING KITCHEN	19. RECEPTION
9. STORAGE	20. OFFICE
10. MAINTENANCE	21. KITCHEN
11. LOADING DOCK	22. BOARDROOM

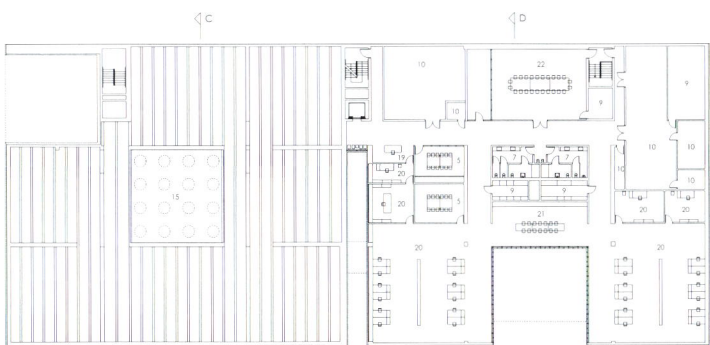

THIRD FLOOR   0' 10'   50'

1. LOBBY AND RECEPTION	12. SECURITY
2. TICKET DESK	13. GALLERY
3. MUSEUM STORE	14. BALCONY
4. MUSEUM CAFE	15. ATRIUM
5. CONFERENCE ROOM	16. WORKSHOP
6. CLASSROOM	17. STAGING
7. RESTROOM	18. REGISTRAR
8. CATERING KITCHEN	19. RECEPTION

The Research Library in Hradec Kralove

Architect: Projektil Architekti
Location: Hradec Kralove, Czech Republic
Area: 13000m²
Photographer: Andrea Lhotáková

The five-storey building represents a concrete construction with the final visual monolithic concrete façade. The original shape of the building is the precast concrete letter "X". From the path one can enter the library, the café and the exhibition hall. In the center of the "X" there is the main vertical communication illuminated by a circle roof light. The distribution of visitors, librarians and books take place from the central vertical point to the four wings of the "X". Public space is located in the two eastern wings and occupies three floors. There are bookshelves, study rooms, individual study rooms and a service counter. In the other wings there are offices, storage and on the 5th floor of office wings is a conference hall. The colored floor creates indoor atmosphere. Inside, there are two wall paintings and a new information system. Important part is the energy–saving concept.

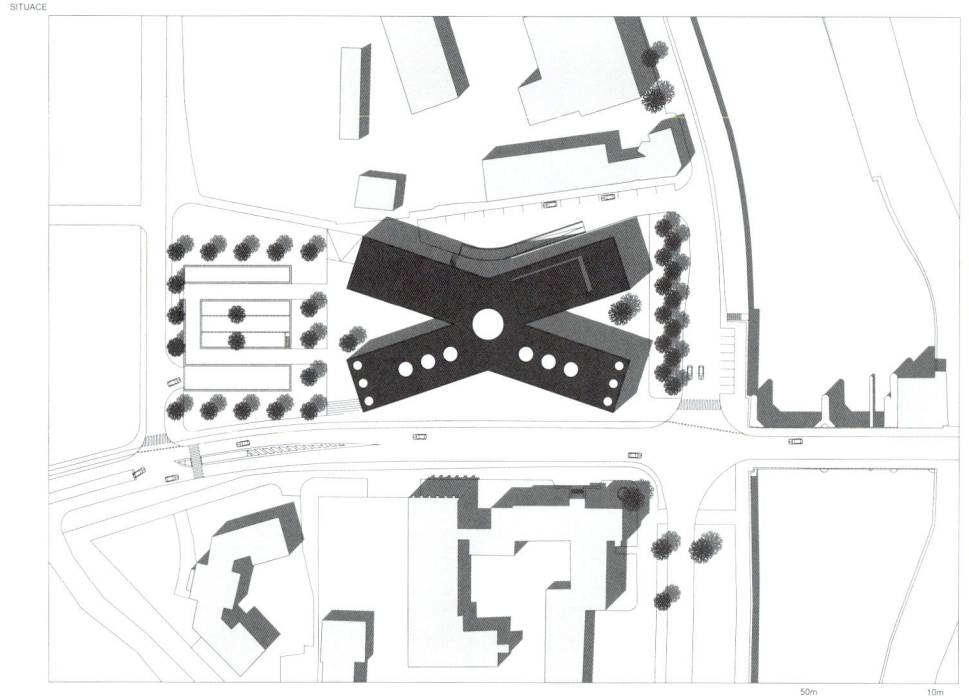

50m   10m

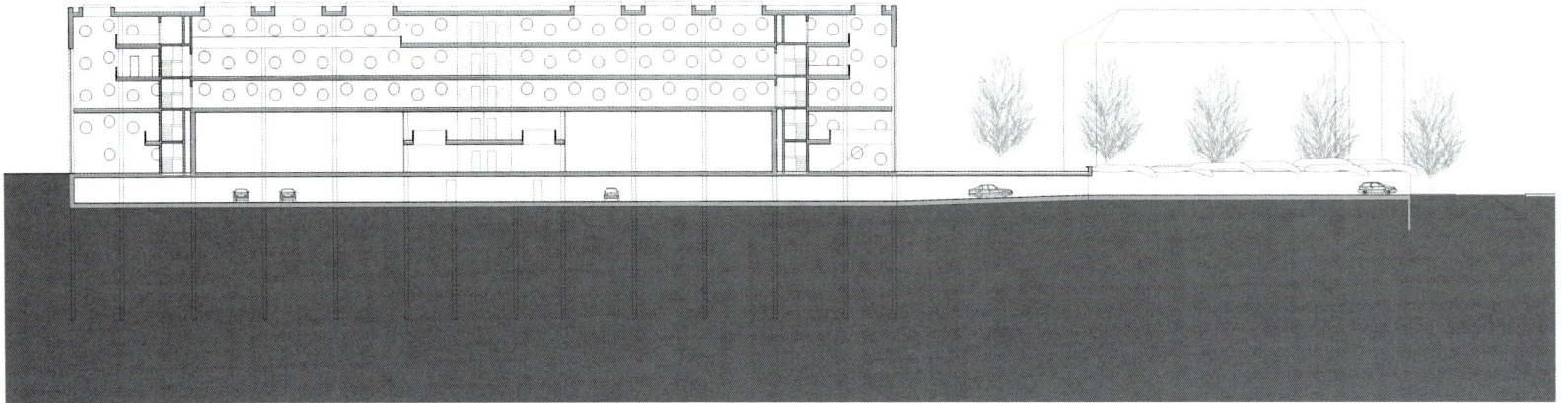

50m   10m

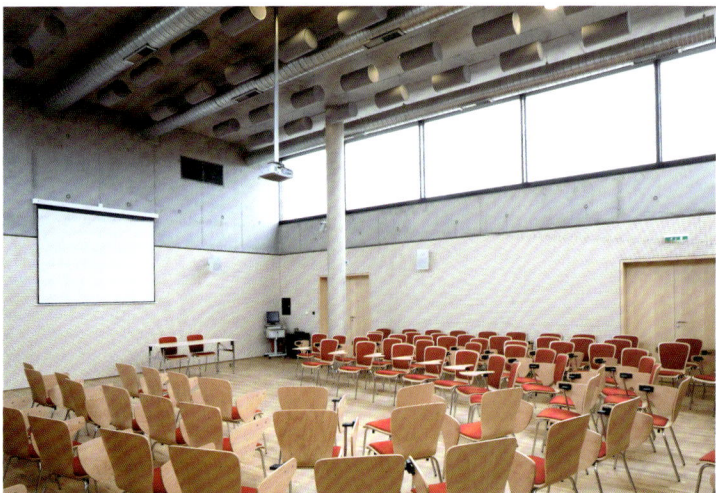

这座五层高的混凝土结构的大楼外观看起来像是一块独立的巨石。大楼是"X"形，你可以从这个字母的底端两划进入图书馆、咖啡馆和展览厅。"X"中心的屋顶，灯排成一个圆圈照射着下面的交流中心。图书馆工作者、书籍和游客分散在"X"的四个方向。东侧是公共区域，共三层，里面有书架、书房、私人书房和一个服务台。另一侧是办公室、仓库，该侧的五楼还有一个会议厅。有色的地板给人一种室内的感觉。图书馆里有两幅壁画和一个信息系统，节能的概念也得到了体现。

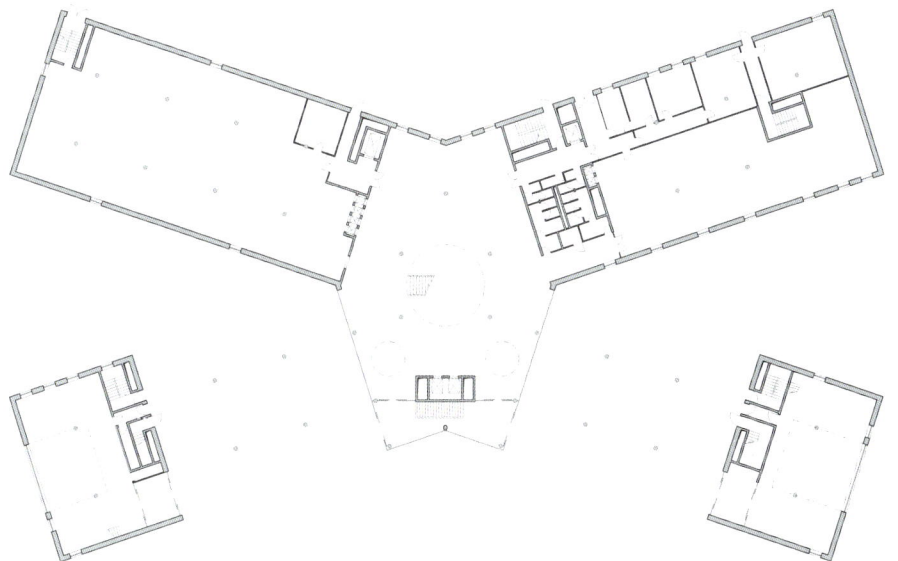

Zenith Amiens

Architect: Doriana & Massimiliano Fuksas
Firm: Studio Fuksas
Loction: France
Area: 11100m²
Material: Membrane, canvas, steel and cement
Photographer: Philippe Ruault

The building was designed as an architectural element like a sculpture autonomous. Adaptation to the needs of the "Zenith", the building also reveals its role emblematic of room for performances and a playful space. The volume is impressive with light using a membrane stretched across the elliptical rings that are screwed around the structure.

Materials will be simple and elegant: canvas, steel and cement for the structure. The exterior finish is a PVC membrane in the mass pigmented red. Coverage below room for performances is a set of hemispheres holes which act as thermal and acoustic insulation. A polyester fabric, covered with PVC "double face" covers the entire building at the same time allowing the spread of a faint red light in the lobby and public areas.

　　设计师把这座大楼设计成为一个类似雕塑的独立建筑。为了满足Zenith公司的需要，设计师把大楼设计为一个象征表演和活动的空间。大楼的外观是用薄膜包裹在螺旋状的椭圆环上，配上灯光效果，让人印象深刻。

　　建筑的材料十分的简单、雅致：帆布、钢筋、水泥。大楼的外墙则用大红色的聚氯乙烯作为表层。

　　在活动空间下面是一组半圆形的小洞，用作保暖和隔音。大楼外墙的材料是涤纶织物外层再涂上聚氯乙烯，这样的双层表面也使得微微发亮的红色灯光可以弥漫在大楼的大厅和公共区域中。

1 ▲

2

4

5

CENTRE EXPO

PLAN DU R+1

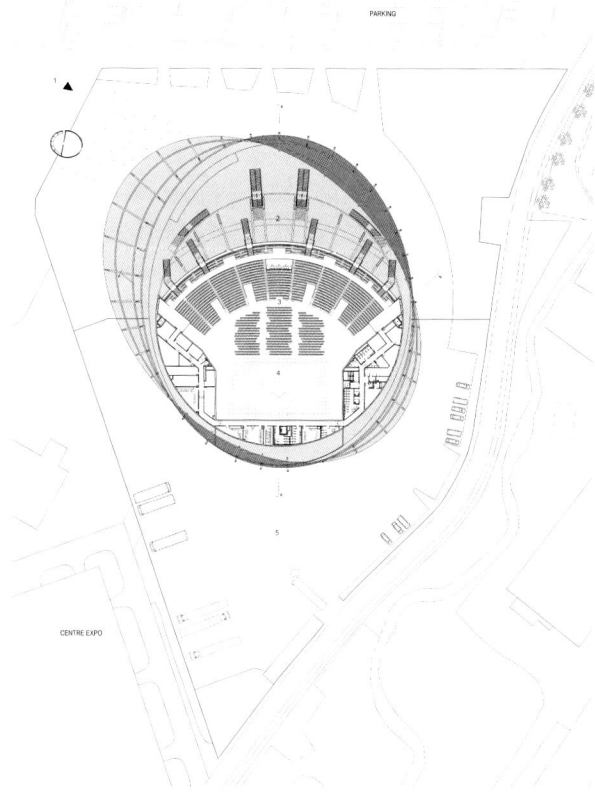

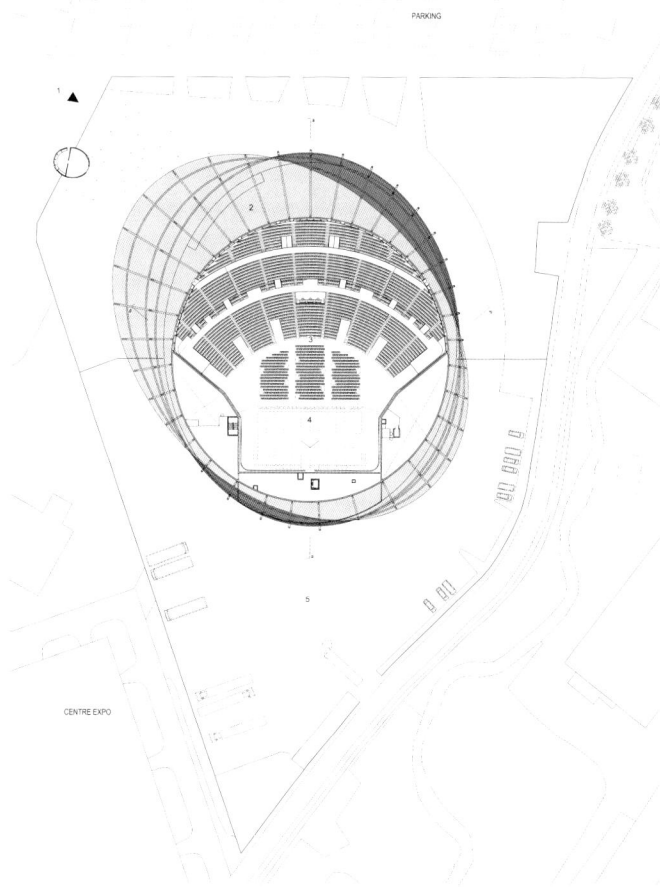

1   ACCES PUBLIC
2   HALL
3   SALLE DE SPECTACLE
4   SCENE
5   PARVIS PROFESSIONNEL

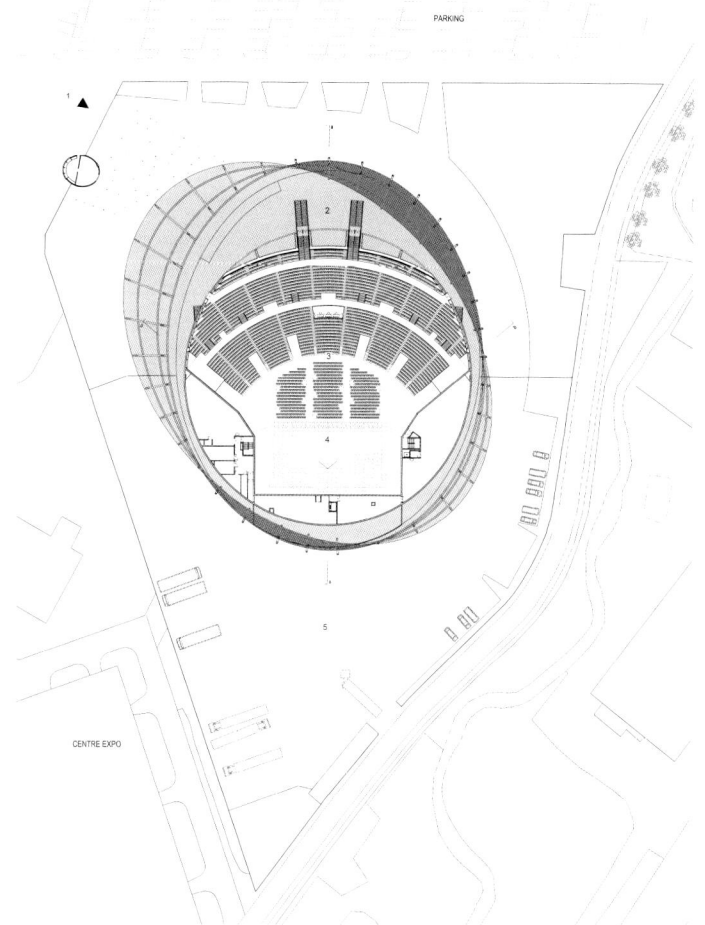

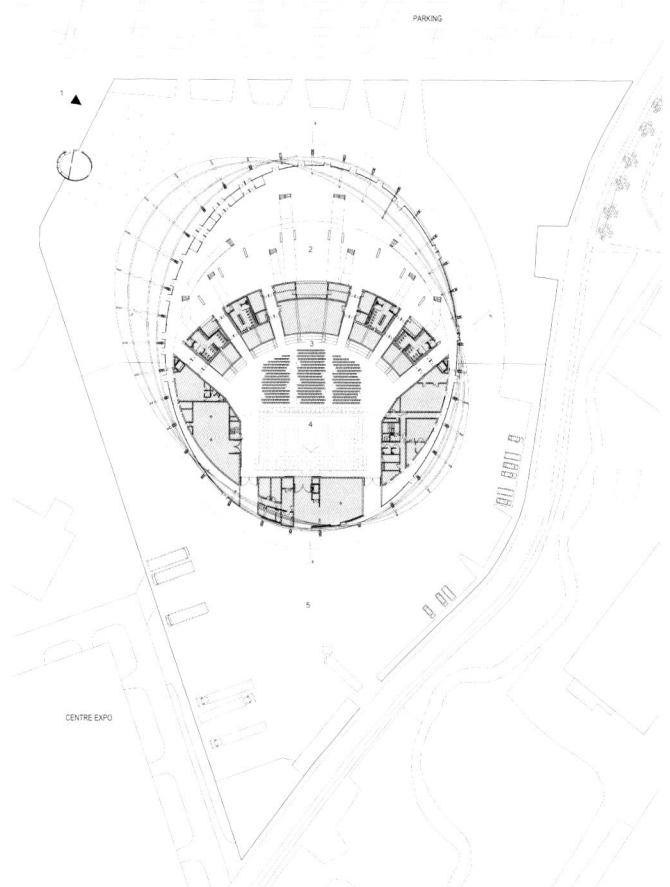

0 5 10   20   50m

1 ▲

2

4

5

CENTRE EXPO

PLAN DU R+2

1   ACCES PUBLIC
2   HALL

0 5 10   20   50m

1 ▲

2

4

5

CENTRE EXPO

PLAN DU R+3

1   ACCES PUBLIC
2   HALL
3   SALLE DE SPECTACLE
4   SCENE
5   PARVIS PROFESSIONNEL

0 5 10   20   50m

1 ▲

2

4

5

CENTRE EXPO

PLAN DU RDC

1   ACCES PUBLIC
2   HALL
3   SALLE DE SPECTACLE
4   SCENE
5   PARVIS PROFESSIONNEL

0 5 10   20   50m

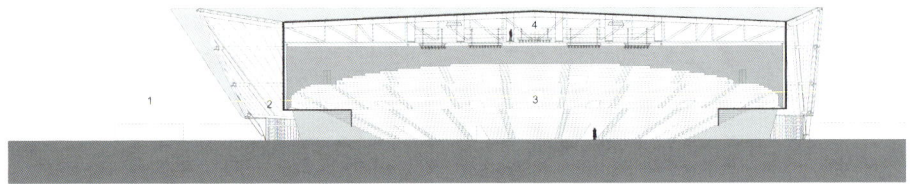

COUPE LONGITUDINALE   b-b

1   ACCES PUBLIC
2   HALL
3   SALLE DE SPECTACLE
4   PASSERELLES TECHNIQUES

0   5   10   20   50m

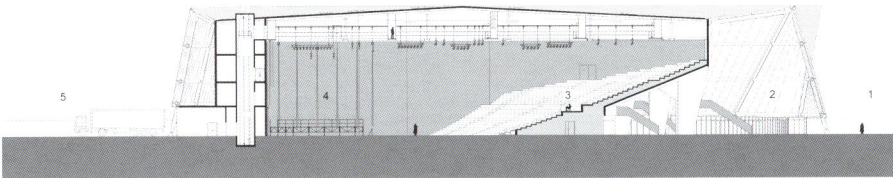

COUPE TRANSVERSALE   a-a

1   ACCES PUBLIC
2   HALL
3   SALLE DE SPECTACLE
4   SCENE
5   PARVIS PROFESSIONNEL

0   5   10   20   50m

Dendermonde Library + Social Restaurant + Multifunctional Space

Architect: Goedele Desmet, Ivo Vanhamme, Jean-Michel Culas
Firm: BOB361 Architects
Location: Dendermonde, Belgium
Area: 6300m²

The building site is situated between the main road of the city of Dendermonde and the green bank of the river Dender. Three urban strategies lead to the creation of a new connection of these two contrasting atmospheres. The historical fortifications are transformed into a greenbelt around the center of the city. The green passage along the library completes the missing link in this structure.

Within the urban fabric the complex introduces a new hangout-place, enriching the social network. The presence of schools and the diverse programs will activate the site day and night. By creating a physical pedestrian walkthrough, connecting the main shopping axis and the recreational green area, a mass of people can penetrate the site.

The passage gives the building four active façades. The folded concrete slab incorporates the lighting and the noise.

Since the building is very deep, the design of a sustainable lighting system was very important: the concrete roof is folded in such a way as to allow natural light to penetrate into the core of the building, without causing glare. To realize optimal summer comfort, intensive night ventilation was integrated in the project. Although library is a very busy and vibrant space, the lighting and acoustic design make it a very comfortable and quiet space.

SECTION AA'

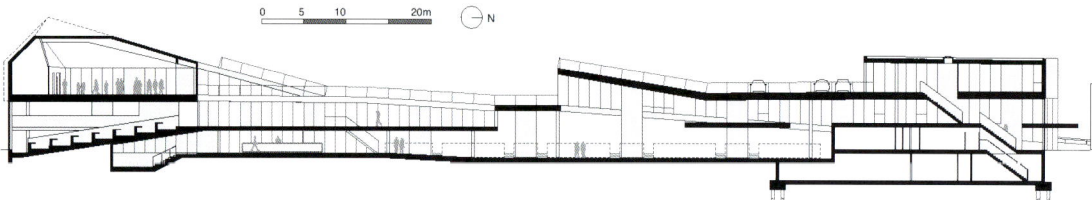

SECTION BB'

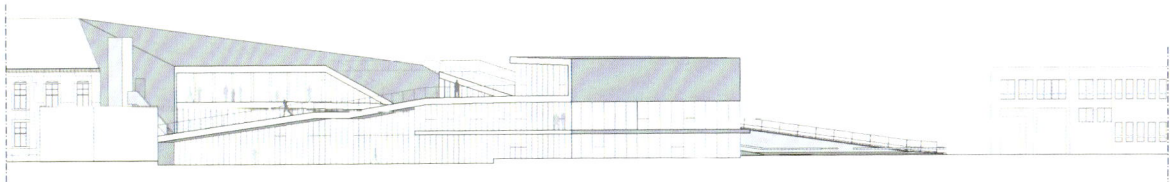

SOUTH FACADE

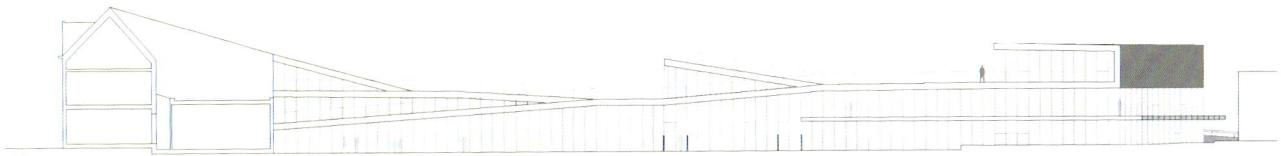

Kerkstraat

WEST FACADE

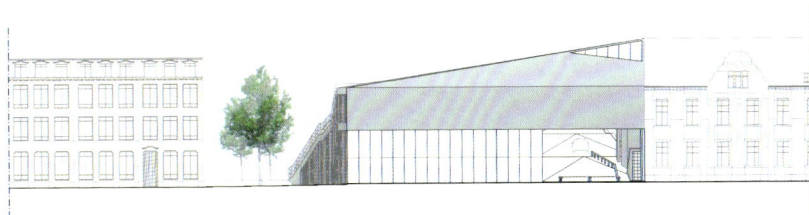

NORTH FACADE (kerkstraat)

BOB361 ARCHITECTEN
LIBRARY AND SOCIAL RESTAURANT
BELGIUM, DENDERMONDE

0   2.5   5   10m

EAST FACADE (passage)

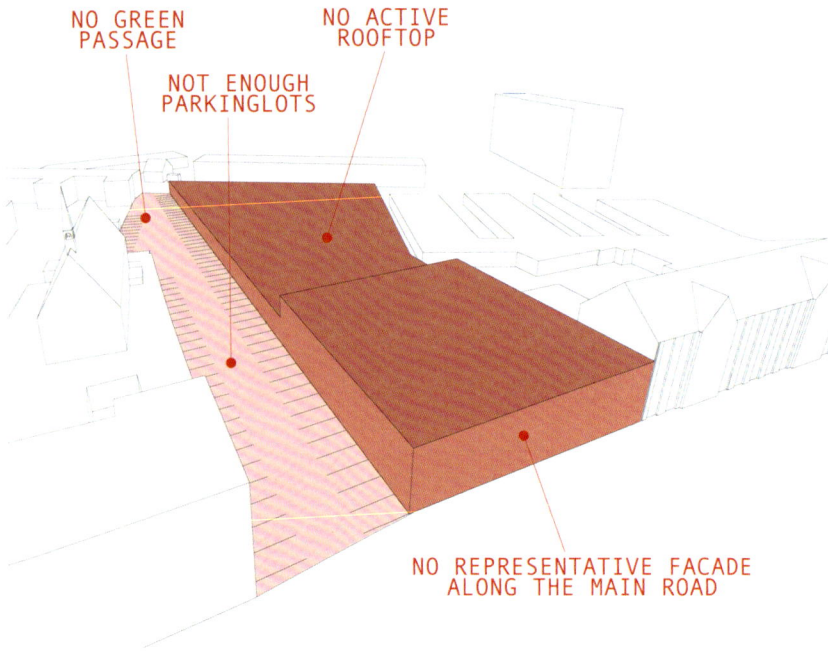

NO GREEN PASSAGE

NOT ENOUGH PARKINGLOTS

NO ACTIVE ROOFTOP

NO REPRESENTATIVE FACADE ALONG THE MAIN ROAD

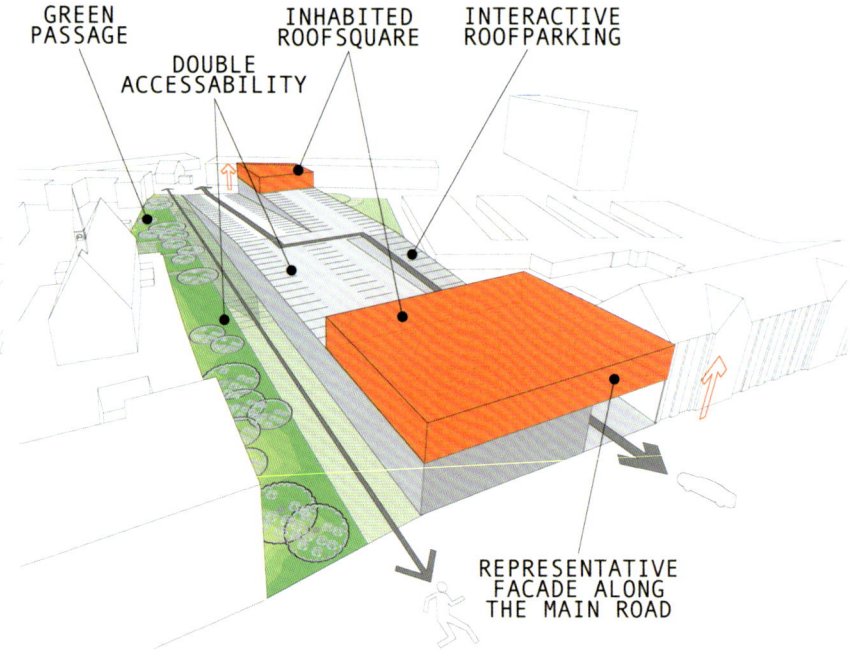

GREEN PASSAGE

DOUBLE ACCESSABILITY

INHABITED ROOFSQUARE

INTERACTIVE ROOFPARKING

REPRESENTATIVE FACADE ALONG THE MAIN ROAD

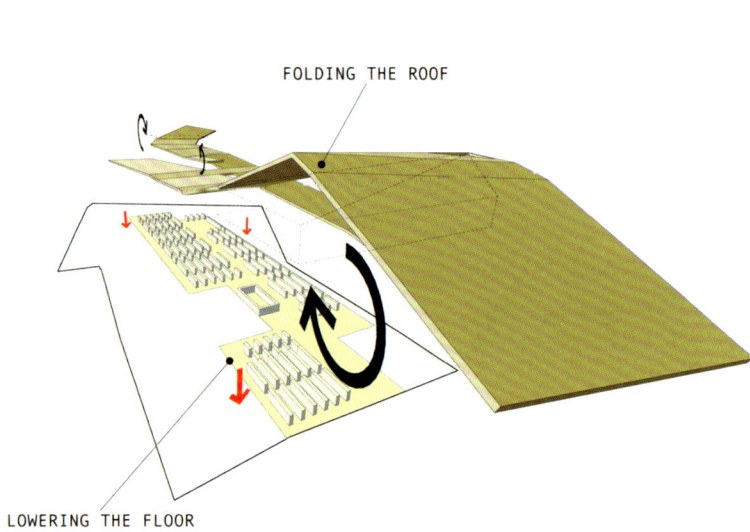

FOLDING THE ROOF

LOWERING THE FLOOR

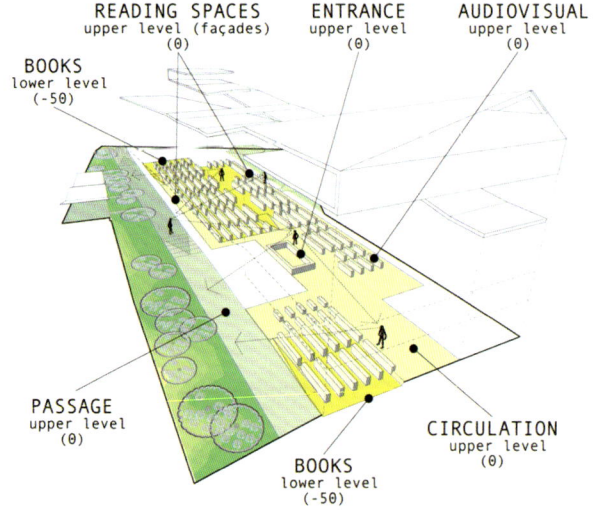

READING SPACES upper level (façades) (θ)

BOOKS lower level (-5θ)

ENTRANCE upper level (θ)

AUDIOVISUAL upper level (θ)

PASSAGE upper level (θ)

BOOKS lower level (-5θ)

CIRCULATION upper level (θ)

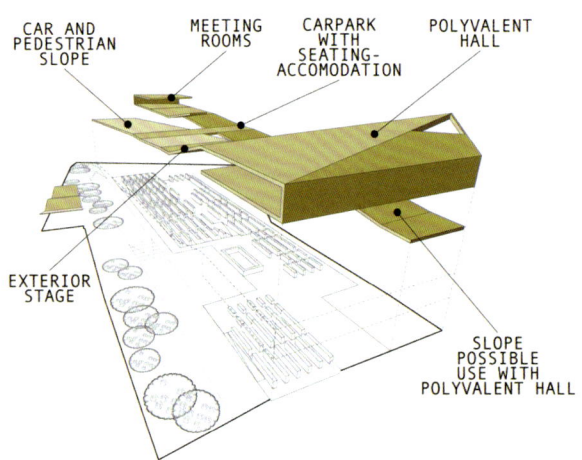

CAR AND PEDESTRIAN SLOPE

MEETING ROOMS

CARPARK WITH SEATING-ACCOMODATION

POLYVALENT HALL

EXTERIOR STAGE

SLOPE POSSIBLE USE WITH POLYVALENT HALL

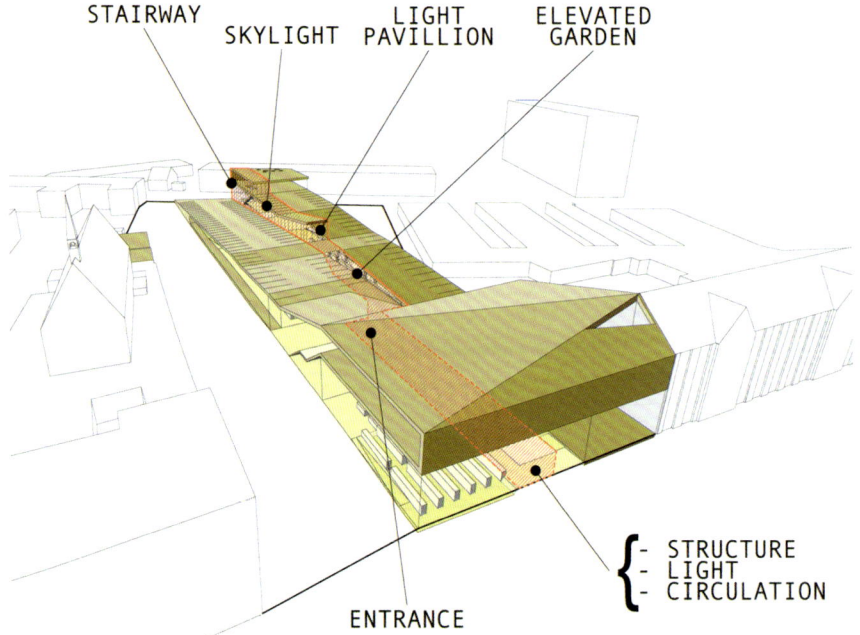

STAIRWAY

SKYLIGHT

LIGHT PAVILLION

ELEVATED GARDEN

ENTRANCE

{ - STRUCTURE
- LIGHT
- CIRCULATION

BOB361 ARCHITECTEN
BIBLIOTHEEK EN SOCIAAL RESTAURANT
OOST-VLAANDEREN_STAD DENDERMONDE

0   5   10   20m   N

1 entrance
2 reception
3 books
4 reading spaces
5 audiovisual
6 children
7 listening space
8 social restaurant
9 kitchen
10 staff
11 bar
12 terrace
13 multifunctional hall

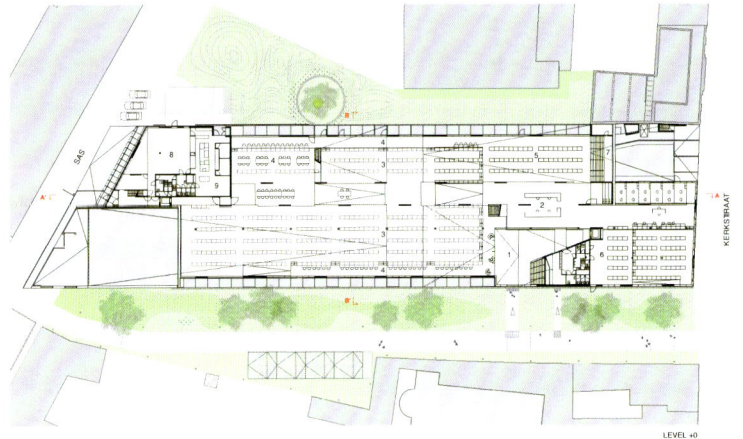

LEVEL +0

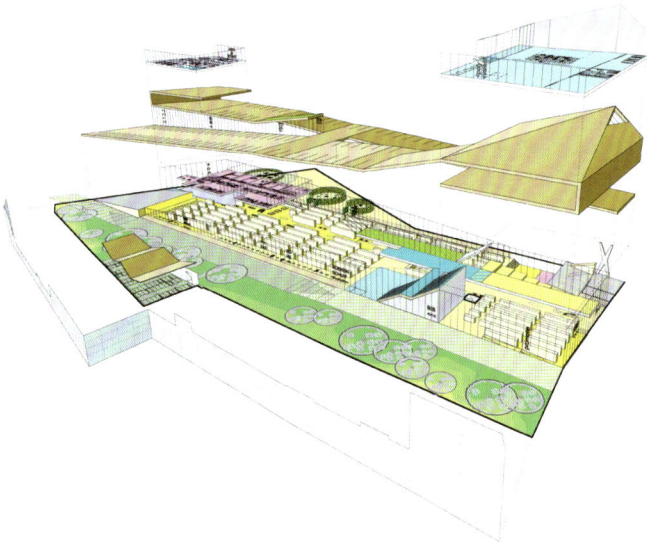

BOB361 ARCHITECTEN
BIBLIOTHEEK EN SOCIAAL RESTAURANT
OOST-VLAANDEREN_STAD DENDERMONDE

0   5   10   20m   N

1 entrance
2 reception
3 books
4 reading spaces
5 audiovisual
6 children
7 listening space
8 social restaurant
9 kitchen
10 staff
11 bar
12 terrace
13 multifunctional hall

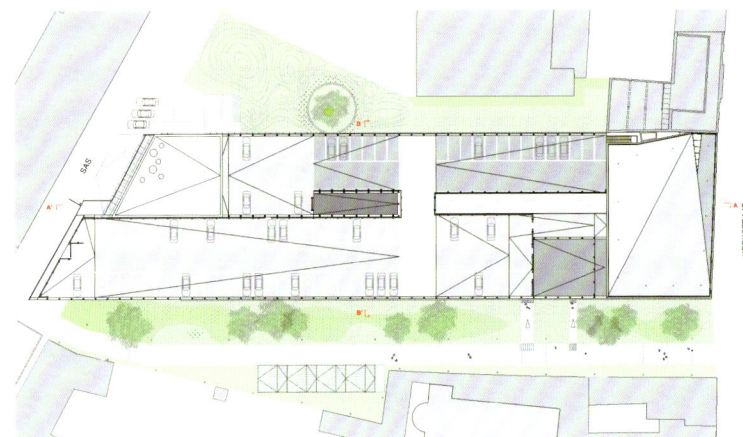

SITUATION PLAN_ROOF LEVEL

该建筑位于Dendermonde市的主干道和Dender河的绿色河岸中间。三个城市战略的出现导致了两个对比鲜明的氛围中一个新连接的建立。曾经的护城墙被改造成一个环绕城市中心的绿化带，这条沿着图书馆的绿色通道弥补了结构中缺少的元素。

在城市结构中，这个综合体提供了一个新的聚集处，丰富了社交网络。学校和其他多种项目的存在将使得这里活力四射。通过一个自然的步行道的设计，将主要的购物轴线和休闲绿化区联系起来，大量的人流可以穿越这里。

这条通道同时也给了建筑4个活跃的立面。折叠的混凝土板吸收了灯光和声音。

由于该建筑的进深很大，所以设计一个可持续的照明系统是非常重要的：混凝土屋顶的折叠方式让自然光渗透到建筑的底层空间且不刺眼。为了实现最佳的热舒适，强化夜间通风设备也集成在项目中。虽然图书馆是一个非常繁忙和充满活力的空间，但灯光和音响设计使它成为一个非常舒适和安静的空间。

National Technical Library

Architect: Projektil Architekti
Location: Prague
Area: 11740m²
Photographer: Andrea Lhotáková

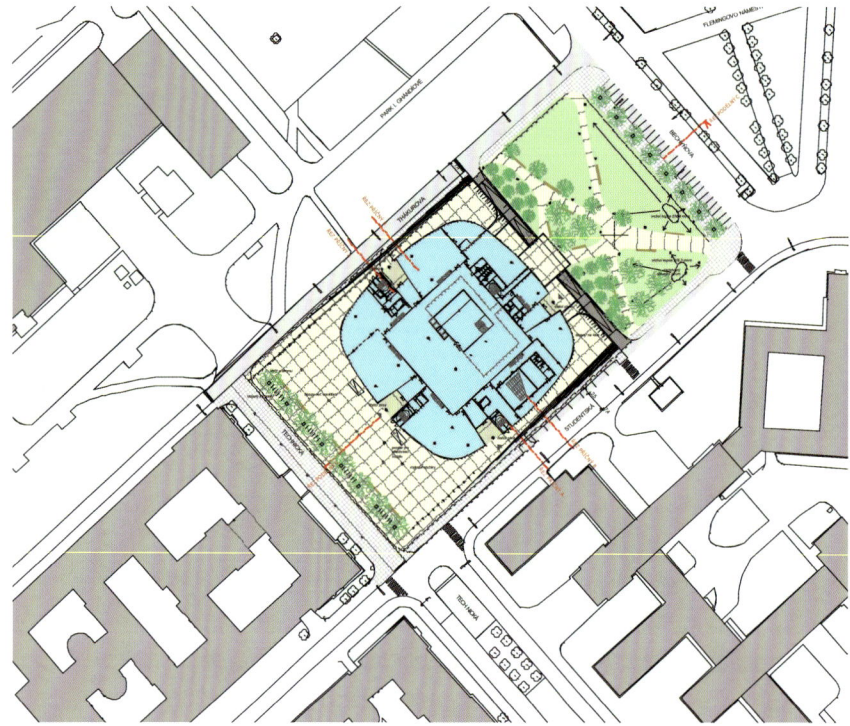

The main principle of interior design is about work together, reciprocal influence.

The main topic of interior design is the colored floor, simply shape of furniture, original design of sofas and the art of the wall.

The idea of the new technical library is quite old and started in the 90's. The architectural competition took place in 2000.

The concept is the answer to the idea of the institution and to the role of the library in today's society. The building should be environmentally friendly.

The entrance to the library with the first info desk is in the middle. The library occupies the upper four floors with the central atrium. Part of the concept is the surrounding area – social space on the west and a green park on the east. The design of the building includes the interior, art and the graphic design following the concept "the technological schoolbook".

GALAXY DNA IPOD
REVOLUTION
SOCIETY
POLITICS
GO PROTEST ON YOUR LEGS
LIFE IS BEAUTIFUL
REFORM
TRAGEDY
STATISTICS
RICH
POOR
ME
WIRELESS
MY WORLD
POWER
b
a

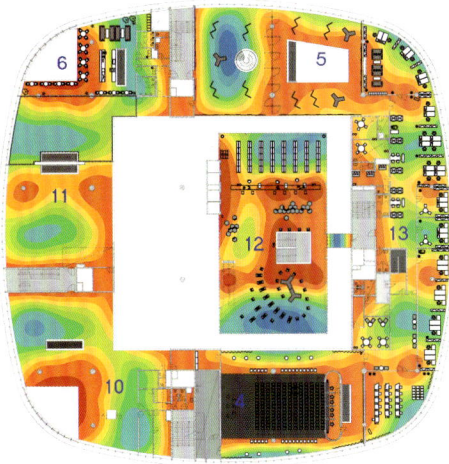

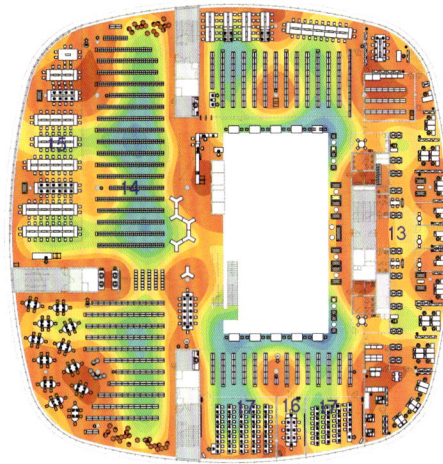

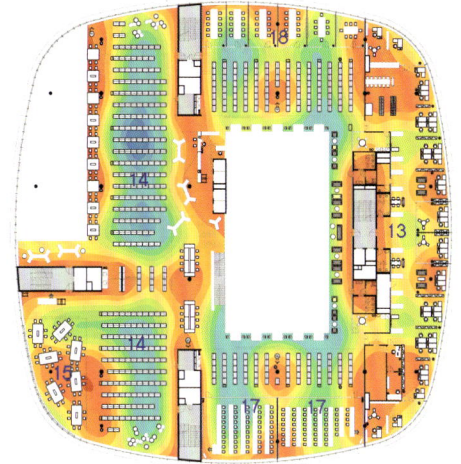

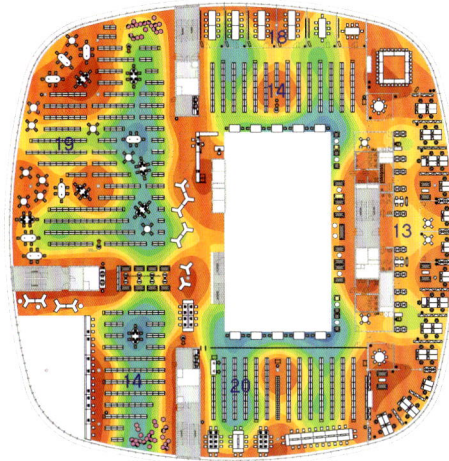

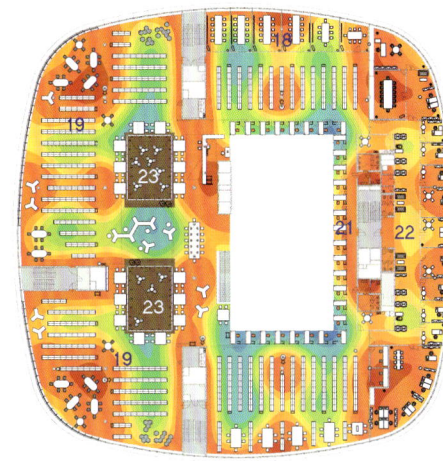

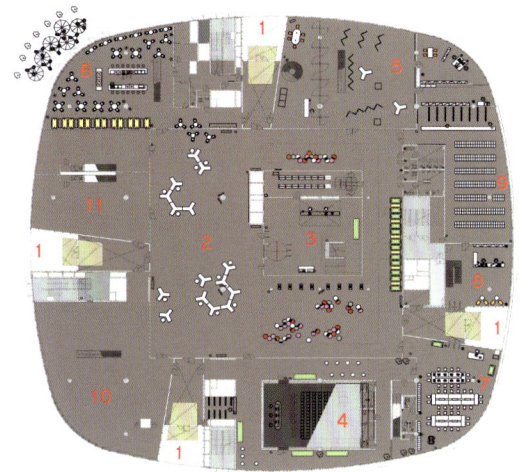

1 - entrance, 2 - lobby, 3 - library entrance, 4 - conference hall, 5 - exhibition hall, 6 - cafe "Technika",
7 - night study room, 8 - staff entry, 9 - cloakroom, 10 - city library (branch), 11 - bookshop, 12 - main library hall,
13 - administration, 14 - bookstore, 15 - study, reading room, 16 - mediatek, 17 - PC room, 18 - team study room,
19 - mixing area, study places and bookshelfs, 20 - standard study room, 21 - individual study rooms,
22 - administration - headquarters, 23 - open atriums

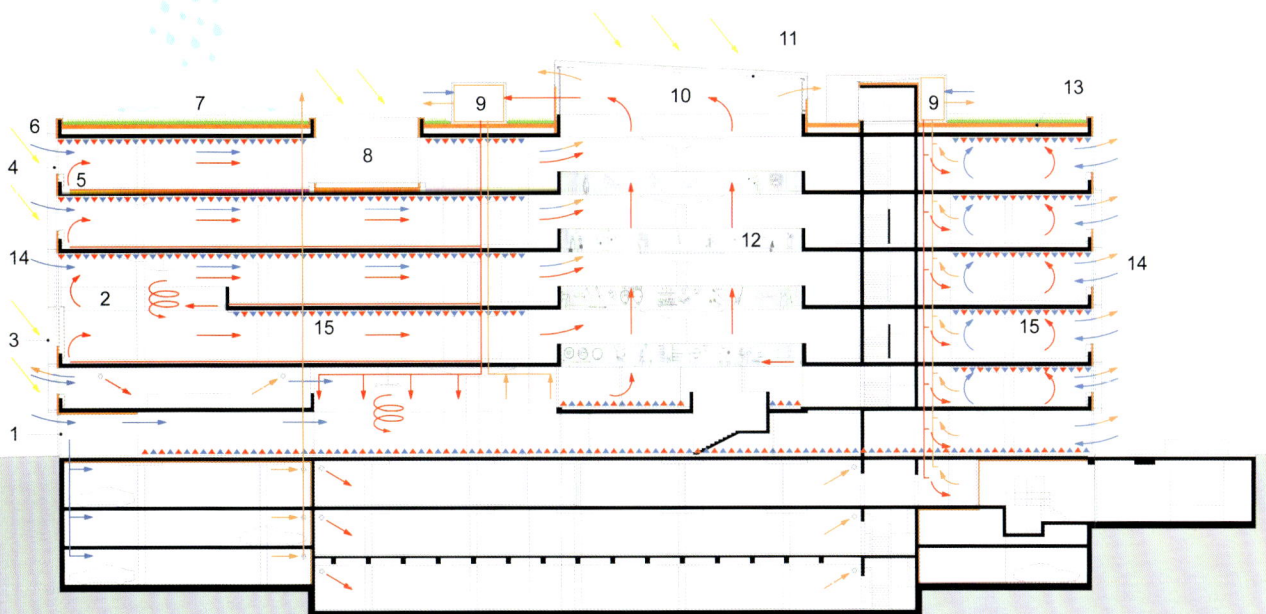

1   mobilní vnitřní roletové stínění snižuje tepelné zisky v
    1np při zabránění kontaktu chodců s ním
2   materiály v přirozené úpravě
3   vnější zasklení z profilovaného skla chrání mobilní
    stínění před větrem a zajišťuje rozptyl světla
4   kontrola solárních zisků - v létě mobilní vnější
    lamelové stínění snižuje tepelné zisky (samostatné
    sekce podle světových stran); v přechodném období
    a zimě solární energie dům vytápí
5   topný registr
6   nasávání čerstvého vzduchu
7   extenzivní zeleň na střeše zpomaluje odtok srážek.
    Chrání izolace a tvoří pěkný vzhled střechy při
    pohledu z okolních domů
8   venkovní atrium pro čtení pod otevřeným nebem
9   vzduchotechnická jednotka s rekuperací tepla z
    odváděného vzduchu
10 zvýšené atrium s prostorem pro akumulaci a
    odvětrání teplého vzduchu v létě
11 vnitřní mobilní stínicí žaluzie
12 kvalitní orientační systém a výtvarné dílo
13 kvalitní tepelná izolace obvodového pláště
14 přirozené větrání zároveň využívané pro automatické
    noční předchlazování betonových konstrukcí v letním
    období
15 topení a chlazení pomocí aktivace betonového jádra

1   internal mobile blind shading declines heat gain in
    1st floor, while avoiding contact with pedestrian
2   materials in their innate condition
3   external glazing made of profiled glass protects
    mobile shading against wind and provides light
    dispersion
4   a solar gain control - in summer external lever
    boards mobile shading causes a decline in heat gain
    (separated sections according to cardinal points);
    during the transmission period and winter the solar
    energy heats the building
5   heating register
6   external air intake
7   extensive planting on a roof for drainage delay,
    protects thermal insulation and makes a nice view of
    roof from surrounding buildings
8   outdoor atrium for reading under open sky
9   a ventilation unit with recuperation of heat from the
    exhaust air
10 elevated atrium with a space for accumulation and
    ventilation of warm air in summer
11 internal mobile blind shading
12 quality orientation system and art piece
13 quality cladding thermal insulation
14 self-ventilation also used for automatic night
    precooling of concrete constructions in summer
    season
15 heating and cooling via concrete core activation

133

室内设计的主要原则就是互相协助、互相影响。

室内设计的主题是有色地板、简约的家私、独特的沙发及艺术感颇强的墙壁。

新科技图书馆的建设项目，早在20世纪90年代就提出了。但建筑设计竞标是在2000年开始的。

这个概念是人们对于图书馆的现代社会角色和本质的回答。图书馆大楼应当设计得环保。

一进入图书馆就可以看到一个咨询台在大厅中间。图书馆共四层，每层中间都有一个中庭圆盘。另外就是周边的环境，该图书馆西边有社区，东边有公园。图书馆的大楼设计包括室内设计、艺术和平面设计等，都是根据图书馆是社会科技教科书这一概念设计的。

Lightcatcher at the Whatcom Museum

Architect: Jim Olson, Stephen Yamada-Heidner, John Kennedy, Olivier Landa, William Franklin, Megan Zimmerman, Michael Picard, Debbie Kennedy, Cristina Acevedo
Firm: Olson Kundig Architects
Location: Bellingham, Washington
Area: 3901 m²

The Lightcatcher at the Whatcom Museum is a regional art and children's museum. It takes its name from its most visible and innovative feature— the lightcatcher— a multi-functional translucent wall that reflects and transmits the Northwest's most precious and ephemeral natural resource, sunlight.

The lightcatcher, 37 feet high and 180 feet long, is at the physical center of the project, gently curving to form a spacious exterior courtyard, while bridging the museum's interior and exterior spaces. During daylight hours, the light-porous wall floods the halls and galleries inside with a warm luminosity, serving as an elegant and energy-saving light fixture.

Outside, the lightcatcher reflects light into the Garden of the Ancients, the 7,000 square feet courtyard designed as a civic gathering space and a dynamic backdrop for sculpture. In the evening, the lightcatcher glows with the colors of the structure's interior illumination. Like a lantern, it provides a warm and welcoming beacon to the community. Pedestrians can view the courtyard—and the art and activities within—through large openings to the street, ensuring the Museum is as active outside as it is inside.

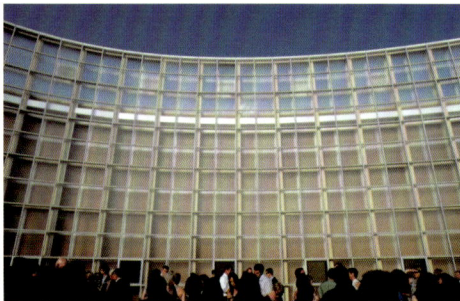

Wall cavity is ventilated to minimize heat gain in the summer. Cool air from the outside flows into the cavity through the bottom louvers and escapes through the top louvers. In the winter the top louvers are closed, the air in the cavity warms up and works as a thermally insulated wall.

Operable louvers allow warm air from the wall cavity to escape. Louvers are regulated by the Buidling Control System that monitors the temperature.

Warm air is able to escape through operable louvers in the roof relief hood if the temperatures rise inside. The Building Control System monitors inside and outside temperatures and closes louvers and windows to maintain a comfortable temperature.

EXTERIOR

INTERIOR

Naturally ventilated space.

Cool outside air flows through operable windows inside. Operable windows are regulated by the Building Control System. They will open when the temperature rises inside and close when the temperature drops.

Hydronic radiant heated and cooled floor.

Open louvers allow outside air to flow into the wall cavity.

Lightcatcher是地域性艺术博物馆的同时还是Whatcom博物馆里的儿童博物馆。这座建筑的名字来源于其突出而极具创意的多功能半透明墙，其墙面能够捕捉美国西北部最宝贵的自然资源——阳光。

高37英尺、长180英尺的lightcatcher是整个工程的核心，它轻轻弯曲形成宽阔的外部庭院的同时，桥接了博物馆的内外空间。白天，透光墙使大厅及画廊阳光满溢，构成生态友好的优雅照明效果。

室外，lightcatcher反射的日光洒满"古代花园"——7000平方英尺的市民娱乐空间及雕刻的动态背景。夜晚，墙面伴随着内部照明色彩的变化而微微发光，像一座给人以温馨和欣喜的指引灯塔。行人可以透过临街的大缺口一览博物馆庭院及馆内艺术品和一切进行的活动，这使得馆外也能分享馆内的活跃气氛。

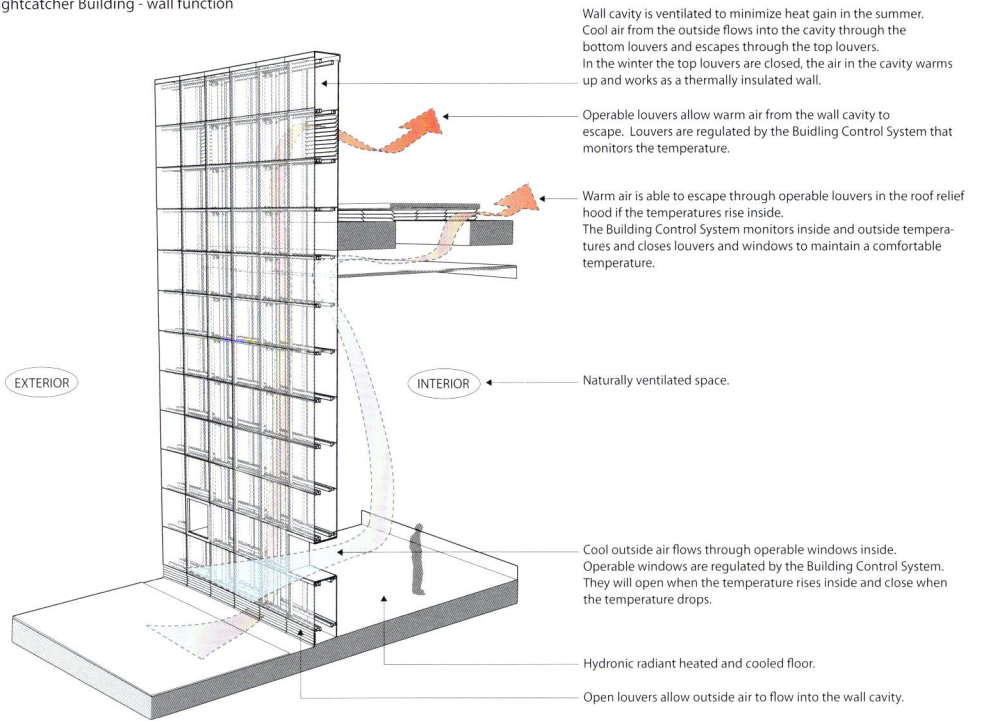

The Lightcatcher Building at the Whatcom Museum

Section diagram with selected sustainable features

1. Green roof and impervious surfacing collect rainwater
2. Detail of naturally ventilated gallery / circulation spaces
3. Cistern to collect rainwater for non-potable building use

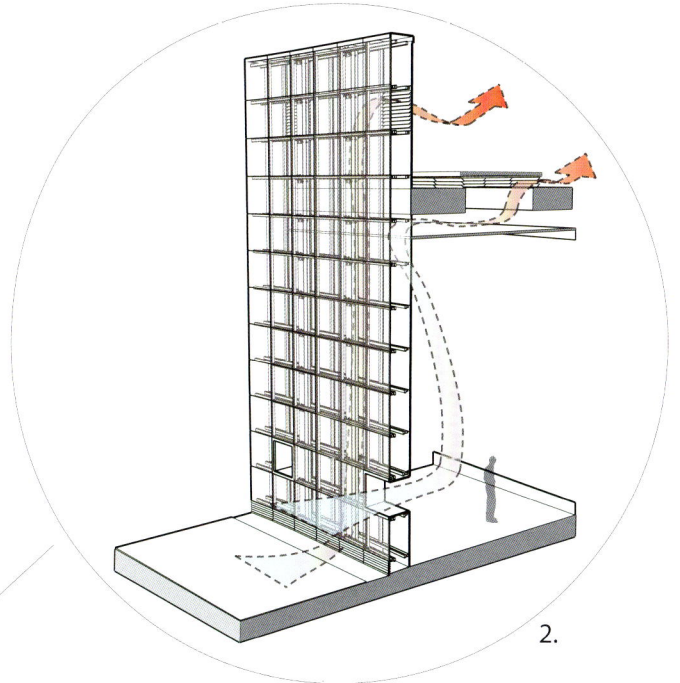

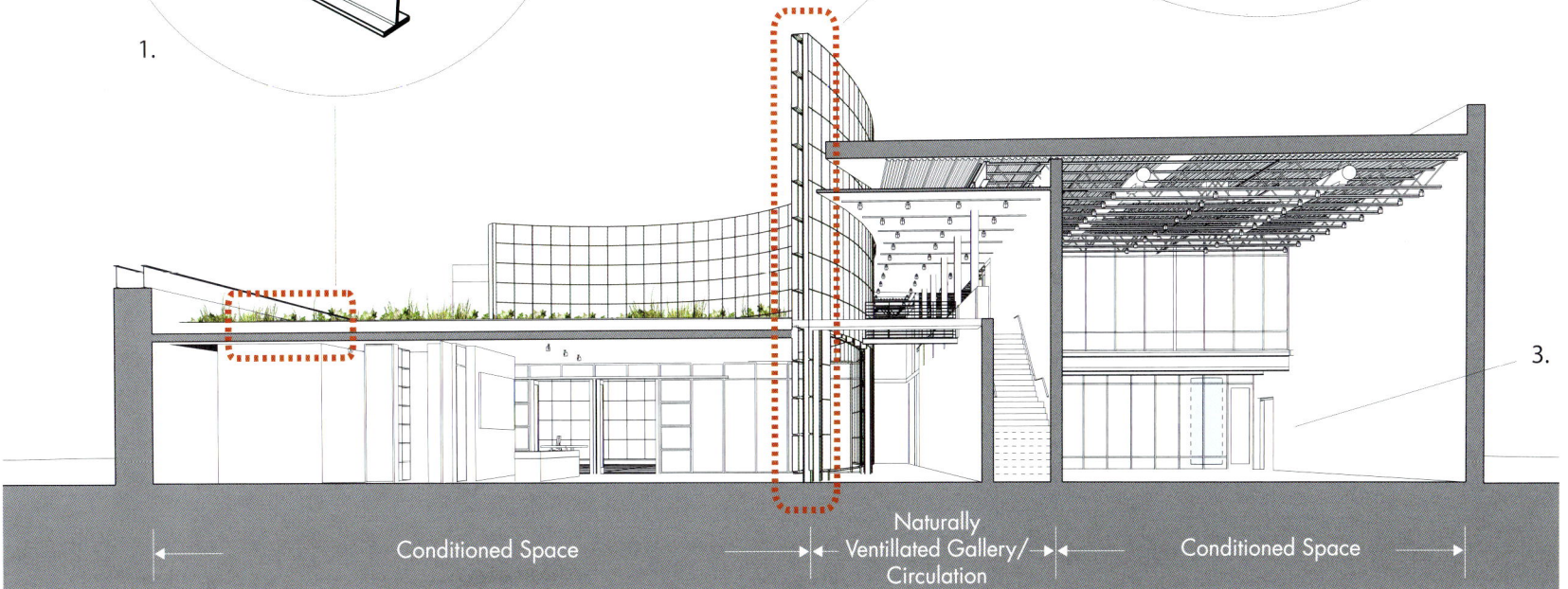

1.

2.

3.

Conditioned Space

Naturally Ventillated Gallery/ Circulation

Conditioned Space

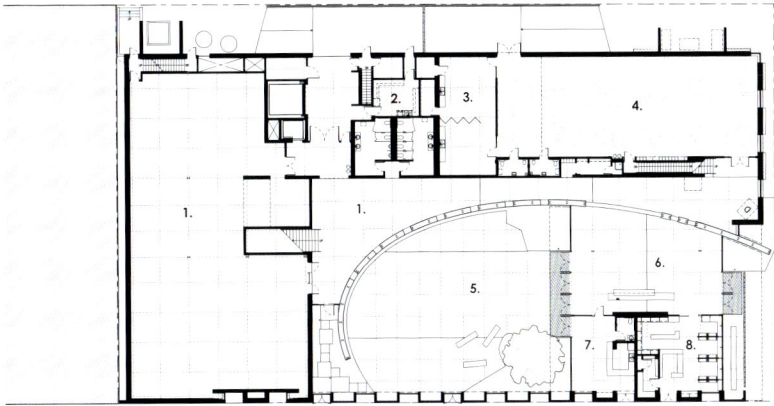

Main floor plan

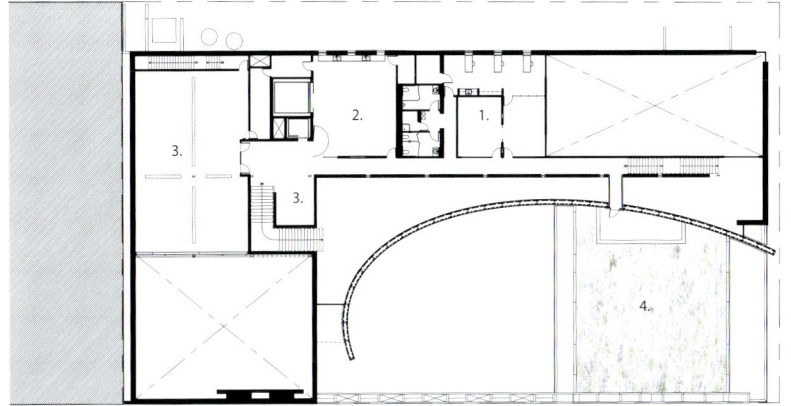

2nd floor plan

Natural History Museum of Los Angeles County

Architect: Scott Kelsey, Jorge de la Cal, Fabian Kremkus
Firm: CO Architects
Location: Los Angeles, USA
Area: 5840m²
Photographer: 2_L Studio LLC, Dick Meir, Tom Bonner

The CO Architects' work at the Natural History Museum began in 2006 with the renovation and seismic upgrade of the iconic 1913 building. The project aims to transform NHM into a state-of-the-art, 21st-century museum.

While changes on the 1913 Building were primarily invisible — the directive was to retrofit and restore the Beaux-Arts gem and its infrastructure without affecting its appearance — CO Architects' reinvention of NHM's North Campus will involve a highly visible transformation of the façades, ingress and egress points, public areas, and exhibition spaces. The redesigned North Campus seeks to enrich the visitor experience, more fully engaging museum-goers with the exhibitions inside, the green spaces outside, and the neighborhood overall, as well as putting an interactive and contextually responsive public face on the museum.

A prominent feature of NHM's new 'front yard' will be a pedestrian bridge — with an arc like the shape of a whalebone — leading from the sidewalk to the museum's first level. Kremkus had dual inspirations for the bridge design — the museum's notable fin whale skeleton and his native Frisian Islands in Germany, where retired sea captains would place massive whale bones in front of their homes to indicate their eminence in the community.

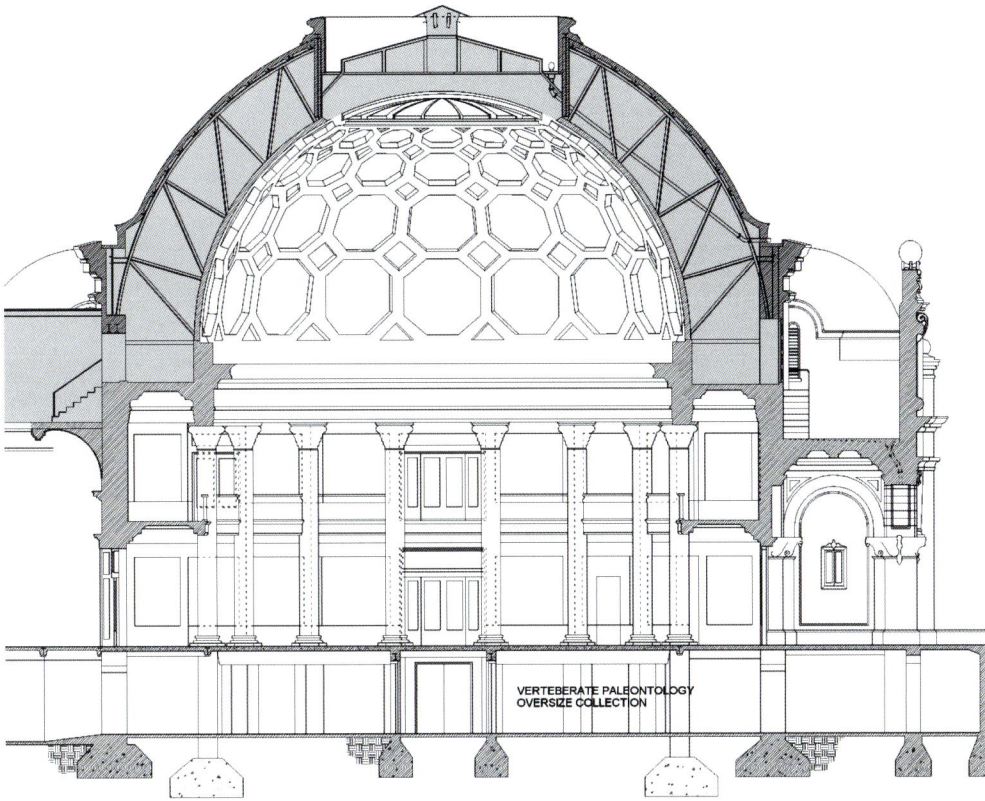

VERTEBERATE PALEONTOLOGY
OVERSIZE COLLECTION

CO Architects建筑公司于2006年修缮和改造了早在1913年就建成的洛杉矶自然历史博物馆，这个项目的目标在于把这个博物馆改造成21世纪的艺术博物馆。

修复必须是无形的，关键在于翻新基础设施的同时要让建筑的外观保留1913年时的模样，以免破坏它的艺术性。北馆将会翻新外墙、进出口、公共区域和展览空间。重新设计的北馆丰富了游客的访问内容，馆内和馆外的绿化空间以及四周都吸引了更多的游人前往，也为这个博物馆带来互动活跃的新面貌。

该博物馆前院的一个突出特征是它的人行天桥，天桥的外形就像一个鲸须，从人行道延伸到博物馆的一楼。这个设计有着双重的意义，其一是代表了博物馆著名的长须鲸标本，另一个是代表德国的弗里西亚群岛，当地退休的船长会把大量的鲸骨头放在家门前来显示他们家族的显赫地位。

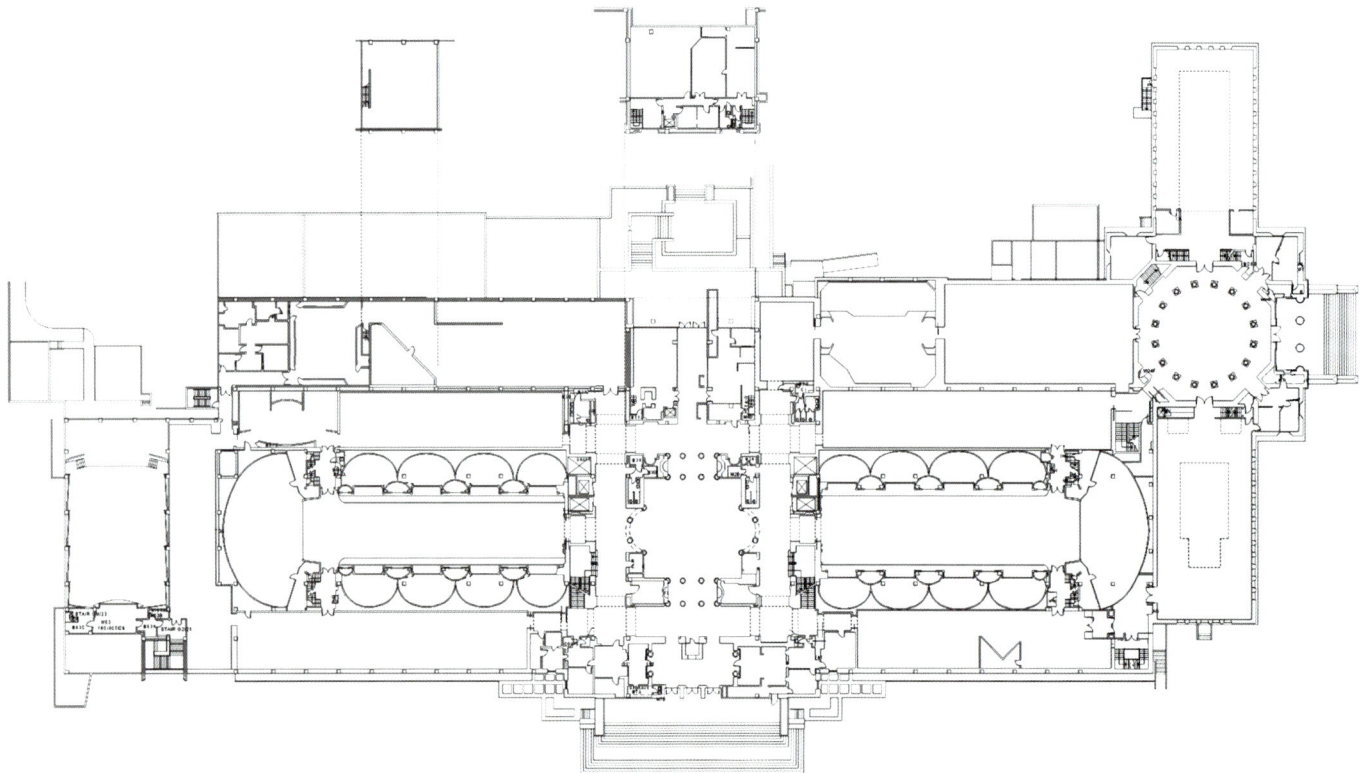

Congregation Beth Sholom Synagogue

Architect: Stanley Saitowitz, Neil Kaye, Markus Bischoff, John Winder, Derrick Chan
Firm: Stanley Saitowitz / Natoma Architects
Location: 301 14th Avenue, San Francisco, CA
Area: 2694m²
Photographer: Rien van Rijthoven (415-810-0973), Bruce Damonte (415-845-6919)

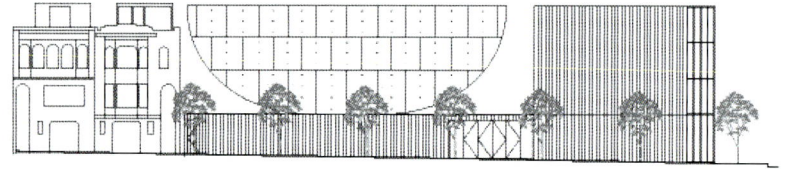

The design for the sanctuary begins from the inside with the creation of a sacred room, a space in the round, focused on the central Bimah from where the services are conducted. A slice of sky in the ceiling turns into the eternal light above the Ark on the Eastern wall. A shadow menorah animates the wall tracing the movement of the sun through the day and illuminated at night. All light enters the room from above with views of the sky creating a sense of sanctity and removing in the midst of the noise and bustle of the city.

The expression of this interior is the exterior of the building. The exterior also remembers the Western wall in Jerusalem, using the color and form of the stones of the ancient temple.

The second building, sheathed in zinc, contains the social hall. It marks the corner with a thin tower like slice. This building, in contrast to the masonry sanctuary, is reflective and light.

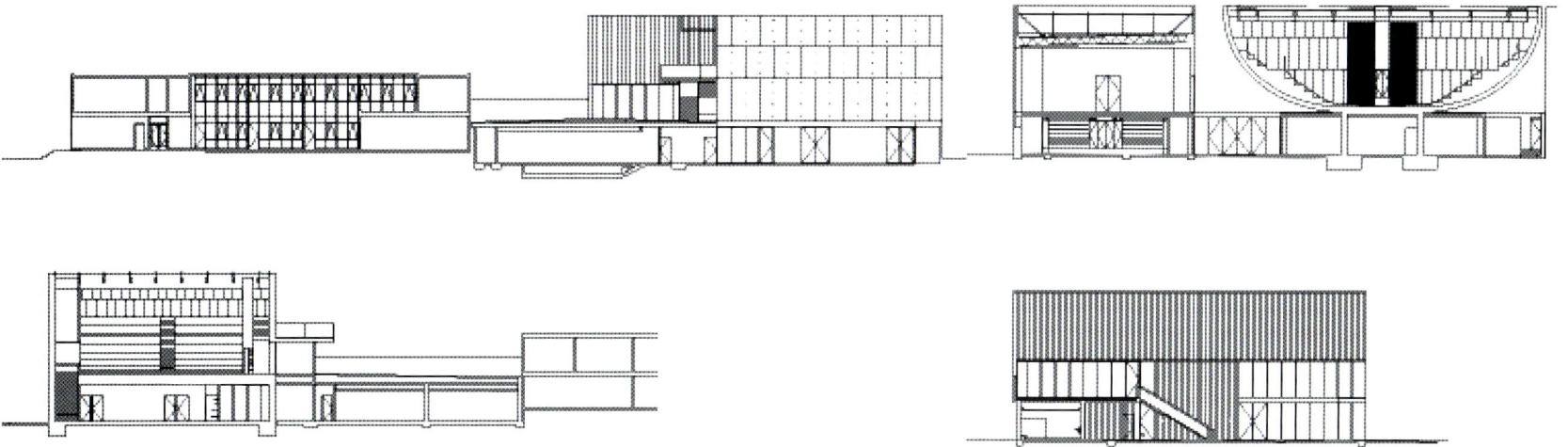

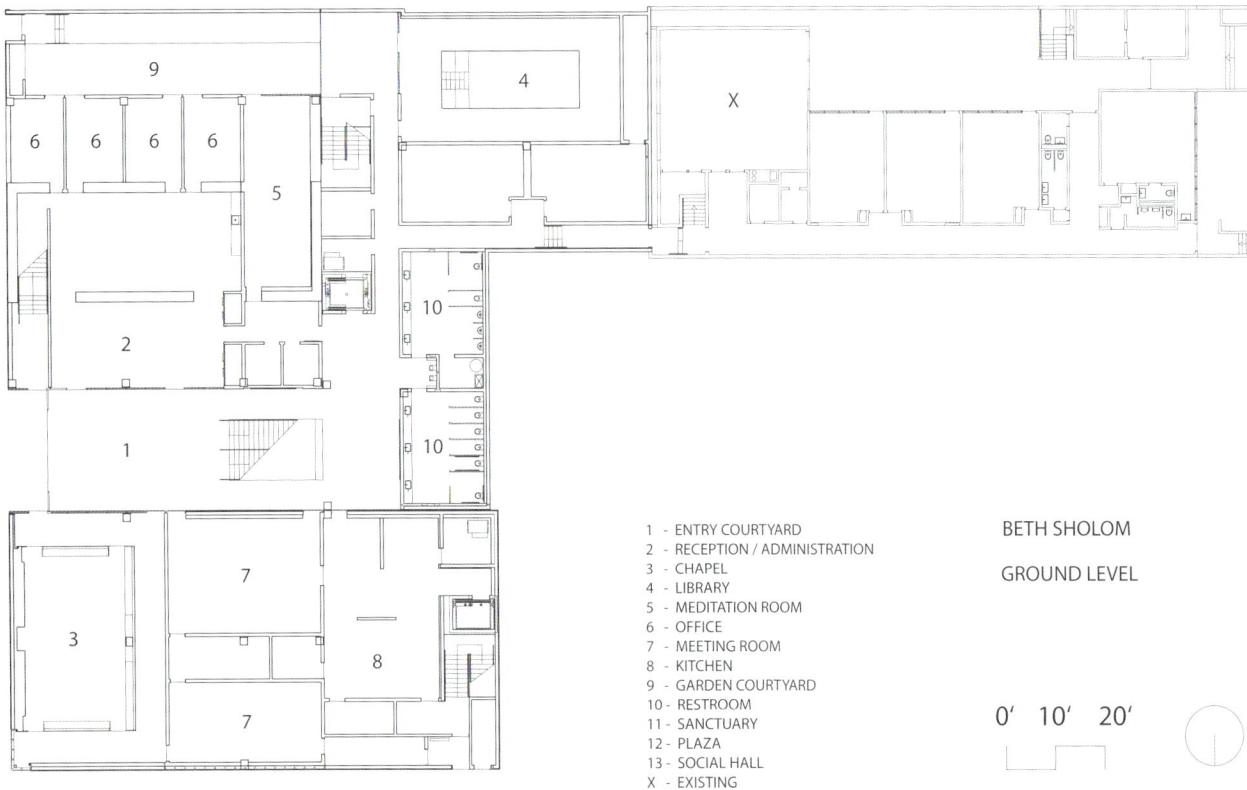

1 - ENTRY COURTYARD
2 - RECEPTION / ADMINISTRATION
3 - CHAPEL
4 - LIBRARY
5 - MEDITATION ROOM
6 - OFFICE
7 - MEETING ROOM
8 - KITCHEN
9 - GARDEN COURTYARD
10 - RESTROOM
11 - SANCTUARY
12 - PLAZA
13 - SOCIAL HALL
X - EXISTING

BETH SHOLOM

GROUND LEVEL

0' 10' 20'

设计师首先设计了一间神圣的圆房，重点设计是在中间的祷告台。在天花板上的一小片天空成为了东墙方舟上面永恒的光。灯台的影子随着太阳的转动而移动，晚上则发出微微的光。所有的灯光从房顶上的那片天空照射进来，带来了神圣，赶走了城市的喧嚣。

室内的设计配合着这个建筑的外在设计。西墙是为纪念耶路撒冷，建筑的外墙使用了古寺石子的形状和颜色。

第二栋建筑外墙的材料为锌，里面是交易厅。它伫立在角落里，犹如一座纤细的尖塔，十分引人注目。与石室祷告房相比起来，这座建筑显得明亮、璀璨一些。

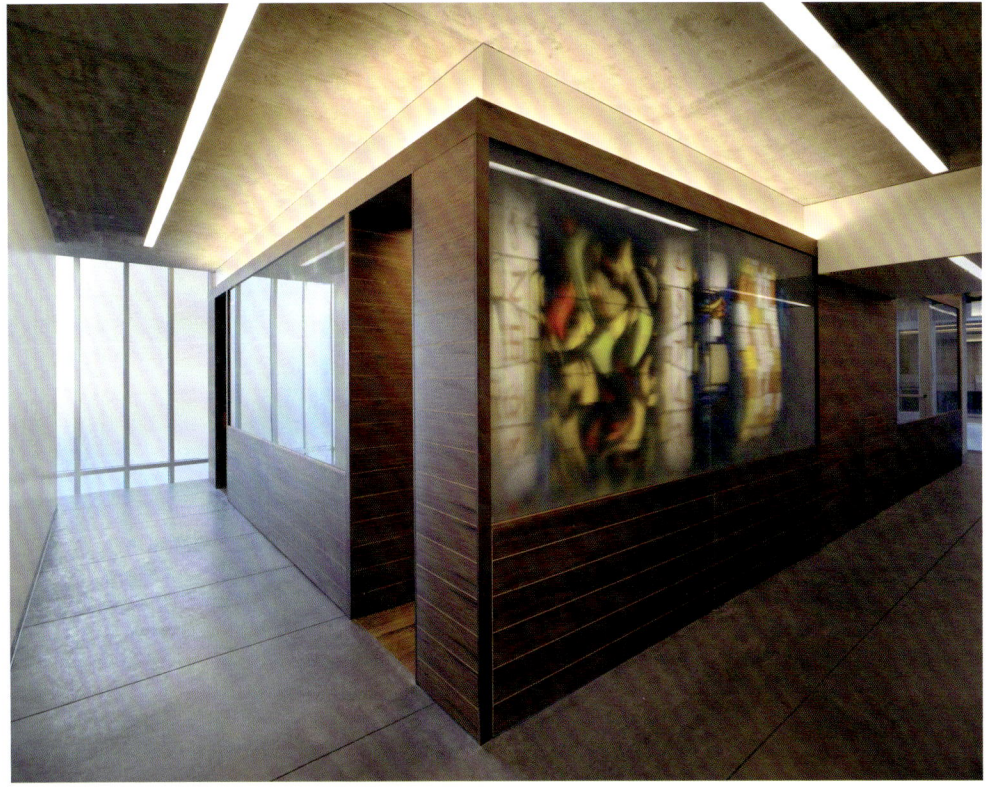

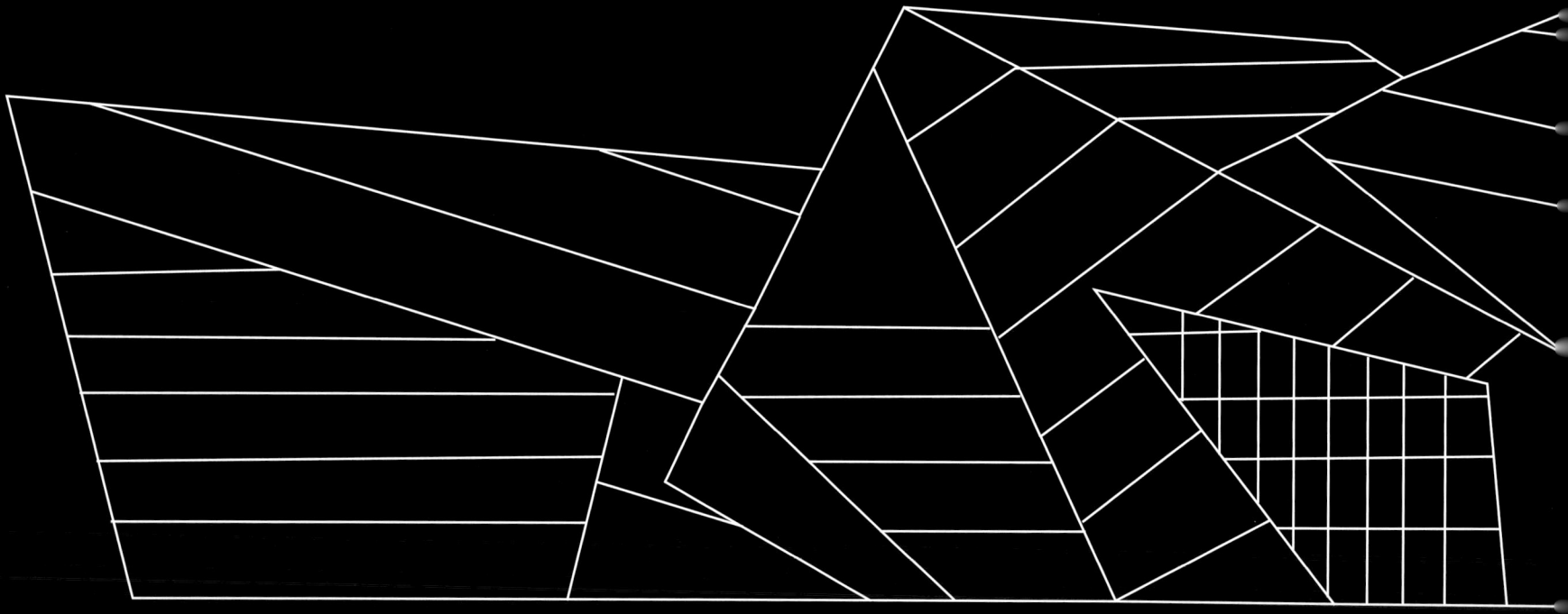

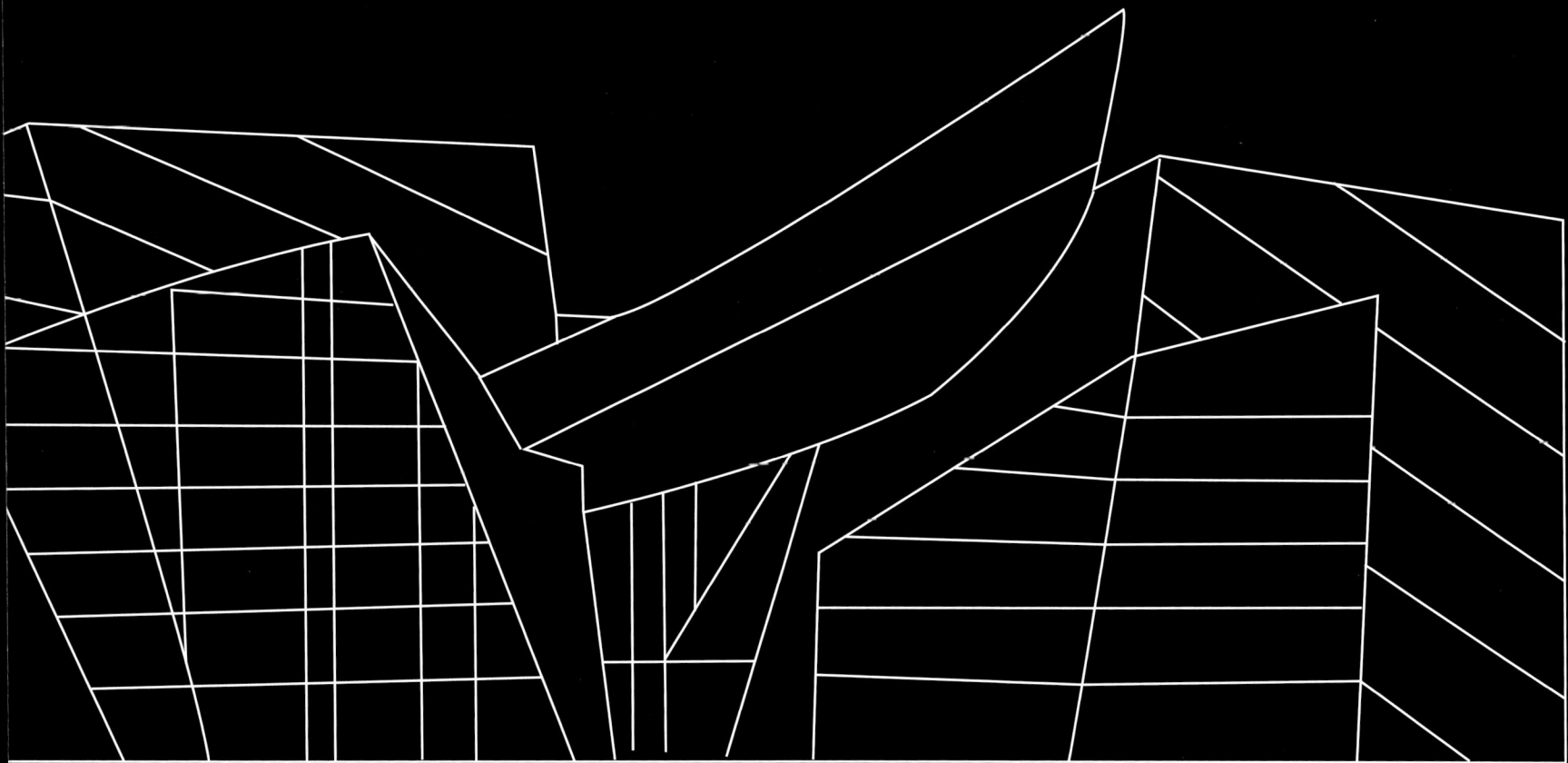

Commercial

商业

MGM Mirage, Las Vegas, Nevada, USA

Architect: Adamson Associates Architects
Location: Las Vegas, Nevada, USA
Area: 46452m²
Structural Engineer: Halcrow Yolles
Facade Consultant: Israel Berger & Associates
Interior Designer: Rockwell Group
Lighting Designer: Focus Lighting
Collaborating Architects: Foster and Partners, Gensler, Murphy Jahn Architects, KPF, Pelli Clarke Pelli Architects, HKS, Leo A. Daly, RV Architecture
Contractor: Perini Building Company
Photographer: CityCenter Land, LLC, SDL, Alexander Garvin, Scott Frances, MGM

Located in the heart of Las Vegas Boulevard, Crystals is the 500,000-square-foot retail and entertainment space, which is the connective center of the MGM MIRAGE City-Center project. Opened in 2009, City Center is a vertical city within a city which includes 2,400 private residences, two boutique hotels, and a 61-story resort casino.

The crystalline and metal-clad facade signals to visitors well in advance of arrival that Crystals is not a traditional retail environment. The passage with a spiral roof leads people to the retail area. From the interior, the roof's dramatic angles and skylights become a backdrop for the luxury retail and dining area which include Louis Vuitton, TIFFANY & CO., and Bulgari as well as concept restaurants from Wolfgang Puck and Todd English. The public spaces allow for a variety of urban experiences: a water feature at the entry, cafes and a grand staircase leading to Casino Square at the end of the arcade, animating the entire space.

The design and construction of Crystals employ the most environmentally conscious practices and materials. In November 2009 it was announced that Crystals achieved LEEDR Gold Core & Shell certification from the U.S. Green Building Council (USGBC), making it the world's largest retail district to receive this level of recognition.

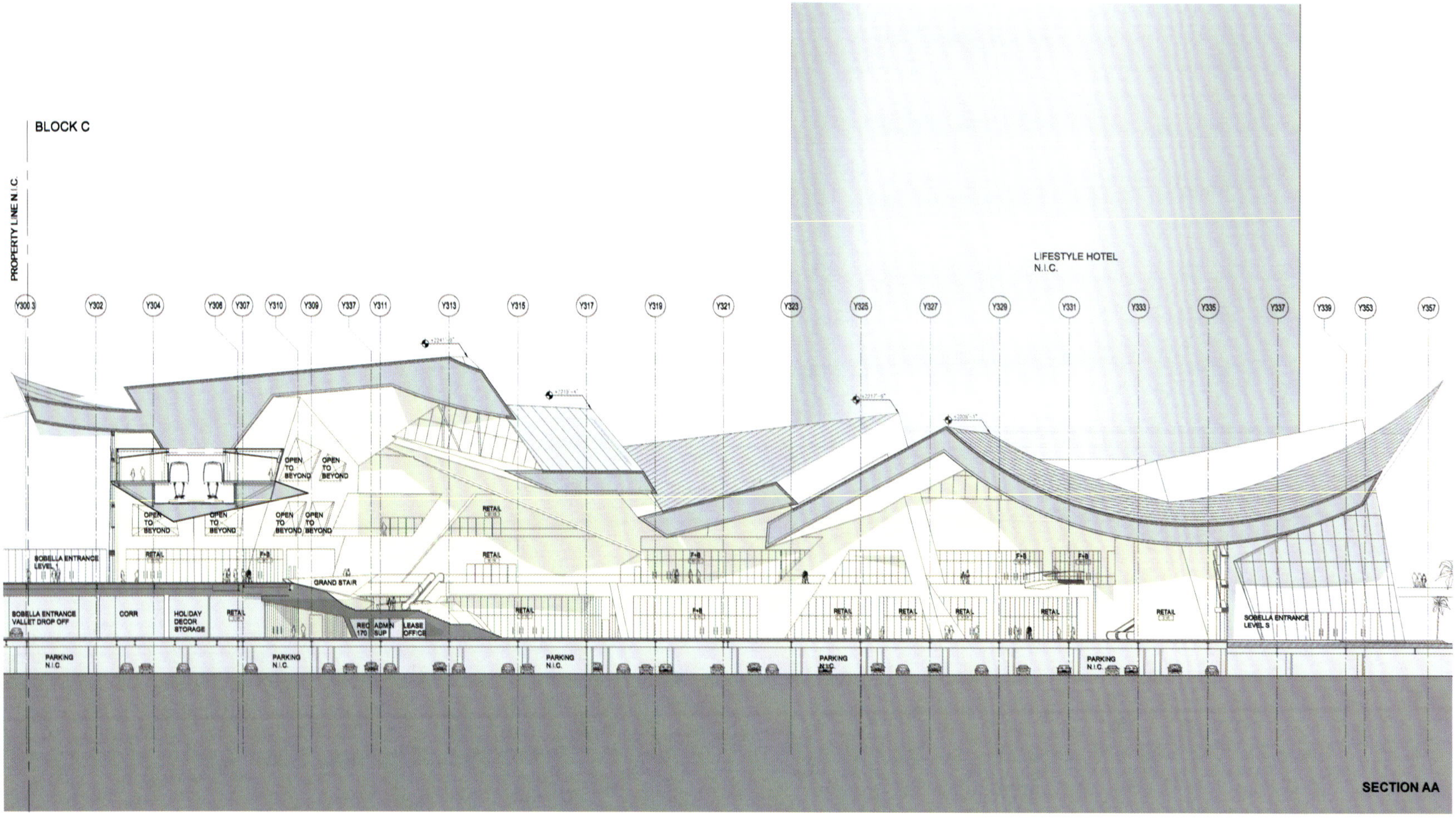

BLOCK C

PROPERTY LINE N.I.C.

LIFESTYLE HOTEL
N.I.C.

Y300.3　Y302　Y304　Y306　Y307　Y310　Y309　Y337　Y311　Y313　Y315　Y317　Y319　Y321　Y323　Y325　Y327　Y329　Y331　Y333　Y335　Y337　Y339　Y353　Y357

OPEN
TO
BEYOND

OPEN
TO
BEYOND

OPEN
TO
BEYOND

OPEN
TO
BEYOND

OPEN
TO
BEYOND

RETAL

RETAL

SOBELLA ENTRANCE
LEVEL 1

RETAL

GRAND STAIR

SOBELLA ENTRANCE
VALLET DROP OFF

CORR

HOLIDAY
DECOR
STORAGE

RETAL

REG ADMN
170 SUP

LEASE
OFFICE

RETAL

RETAL

RETAL

RETAL

RETAL

RETAL

RETAL

SOBELLA ENTRANCE
LEVEL 5

PARKING
N.I.C.

PARKING
N.I.C.

PARKING
N.I.C.

PARKING
N.I.C.

PARKING
N.I.C.

SECTION AA

LOUIS VUITTON

Crystals位于加斯大道的中心地带，是一个占地50万平方英尺的零售和娱乐空间，这是米高梅市中心项目。在2009年开业的城中城里面有2400间私人住宅、两个精品酒店、61层的垂直城市度假村赌场。

建筑的表面材料是晶体和金属，使游客感受到这并不是一个传统的零售商场。一道螺旋形屋顶长廊把游客带进了零售区域，室内，屋顶梦幻般的天窗下是各个零售和餐饮的奢侈品牌，包括路易威登、蒂凡尼、宝格丽以及沃尔夫冈帕克和托德英概念餐厅。这里可以带给你各种都市体验。门口有一道水景，咖啡厅林立，装饰豪华的楼梯带领你来到商场最里边的赌城广场，整个商场生机勃勃。

Crystals环保意识的做法和材料，使它在2009年11月通过了美国绿色建筑委员会（USGBC）LEEDR认证，成为世界上获得这个级别认可的最大的零售区。

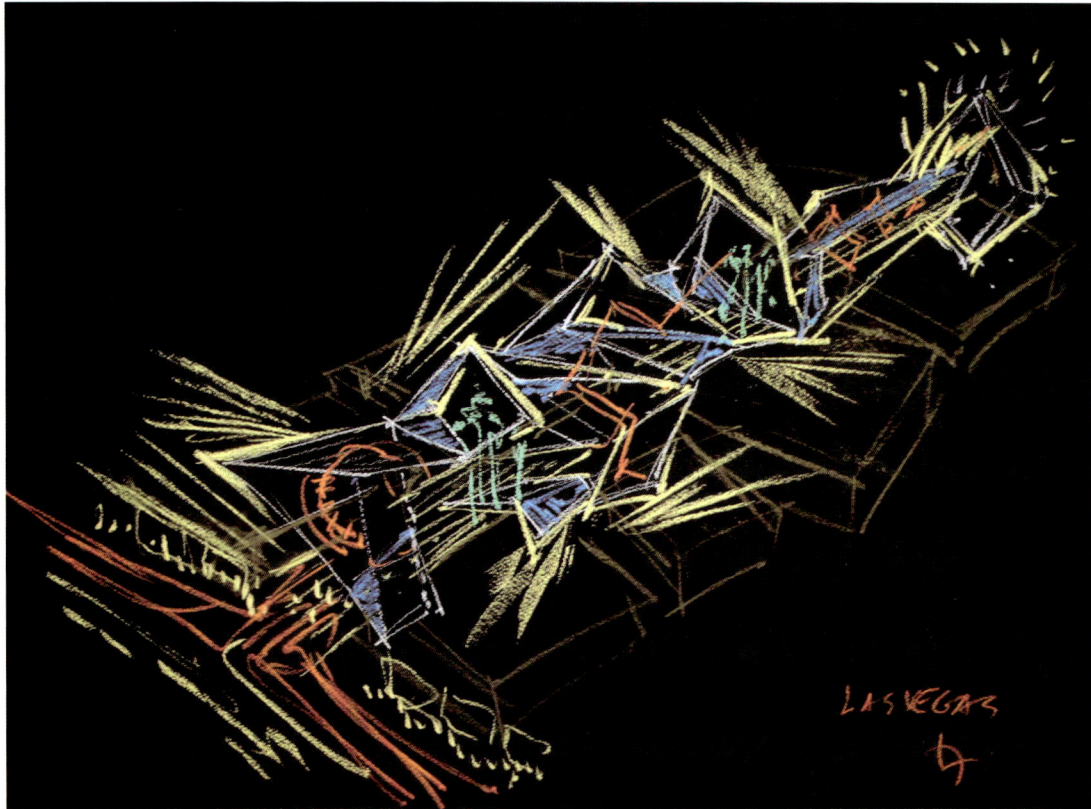

Atrio

Architect: Andrei Florian
Firm: ATP Architects and Engineers, Innsbruck
Location: Villach, Austria
Area: 97710m²
Photographer: Thomas Jantscher, Elke Visciotti

ATRIO in Villach, Austria, was named the world's most sustainable shopping center at the ICSC (International Council of Shopping Centers) "night of the stars" gala during the retail spring convention of the international industry organisation at Caesars Palace.

Completed in 2007, the building provides 38,700 sqm of lettable space and is the first themed shopping center in Austria. The impressive form of the large and powerful stand-alone building dominates the southern access to Villach, greeting visitors like a red and silver sculpture. The architecture responds to this theme of "Borders and the Crossing of Borders" in many ways. An almost square and glazed atrium measuring 50m x 60m has been created which establishes a new urban setting for the city of Villach in the way in which it encourages all such activities of the traditional market place as shopping, eating, playing, communicating and just hanging around.

Multi-cultural design is the end result here, reborn in both the center's name and its central plaza as a focal point. Its design arranges the mall and the INTERSPAR hypermarket around this glazed plaza, effectively creating a new urban square for Villach.

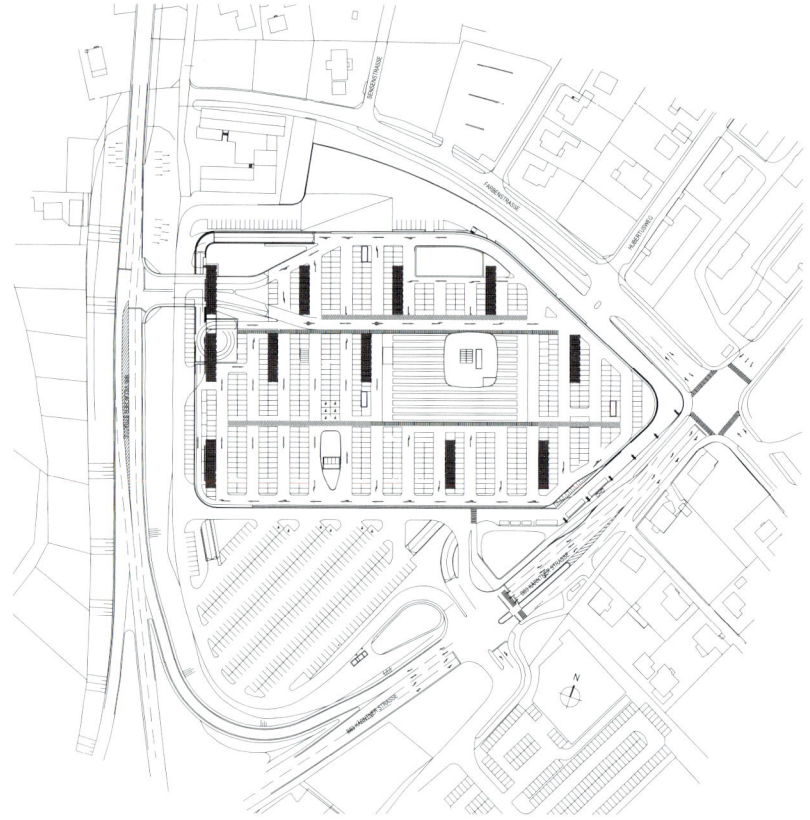

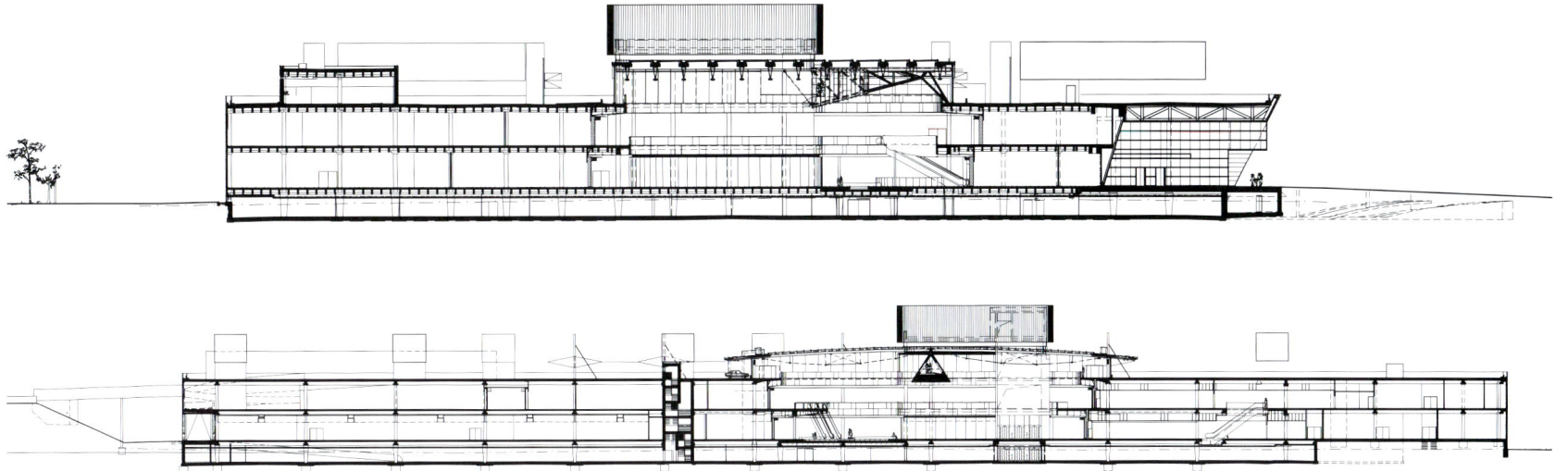

在凯撒皇宫酒店举办的国际行业组织零售行业春季大会期间，位于奥地利菲拉赫的ATRIO，在国际购物中心协会"黑夜之星"的庆典上被评为世界最可持续发展的购物中心。

建成于2007年，该建筑提供了38700平方米的可出租空间，是奥地利的第一个主题购物中心。这栋大而有力的独立建筑令人印象深刻的外形高耸于菲拉赫的南部入口，像一个红色的银雕塑一样欢迎着游客。该建筑在很多方面都响应着"边界和越过边界"的主题。一个50米 × 60米几乎成方形的玻璃中庭广场的创建为菲拉赫市建立了一个新的城市环境，它用一种支持所有传统市场的活动比如购物、吃、玩、交流和只是闲逛等同样的方式建立了这个新环境。

多元文化的设计是这里的归宿，在中心的名称和作为焦点的中心广场这两点上都得到了新的诠释。它的设计很好地安排了这个玻璃广场附近的商场和INTERSPAR大型超市，为菲拉赫有效地创建了一个新的城市广场。

MyZeil

Architect: Doriana & Massimiliano Fuksas
(portrait photographer: Carlo Gavazzeni)
Firm: Studio Fuksas
Location: Frankfurt, Zeil (street), Germany
Area: 78000m²
Photographer: Moreno Maggi

The basic idea of the Frankfurt Hoch Vier shopping mall is to create a vertical city, an urban shopping street with a new element verticality. Physical and visual connections inside the building but also to the urban landscape of Frankfurt integrate this huge project to the urban scale of the inner city of Frankfurt.

The connection of the main facade with the fluid roof, a canyon with voids, is a characteristic element for the project.

The emphasis of the connection builds the important entrance to the main shopping street of Frankfurt, the Zeil. It attracts people to enter the fluent interior space, which connects the Zeil with the Thurn und Taxis palace and the "Frankfurter Rundschau" area on one side, but also the ground floor with the multi-level used retail and amusement center. Entering deeper in the building, void systems that descend from the fourth level, bring down the light and bring up the people. The fourth level which is directly connected with an expressway to the ground floor is becoming a second ground-floor. It's character is a brightly natural lit urban shopping street with an attractive leisure program. Distribution for functions on the fourth floor and the levels above as well as connections to the hotel and the office-building nearby are creating an alive and attracting level.

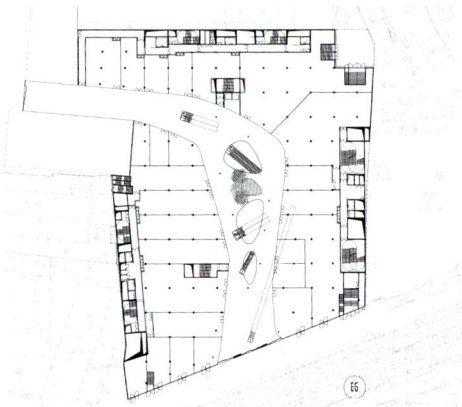

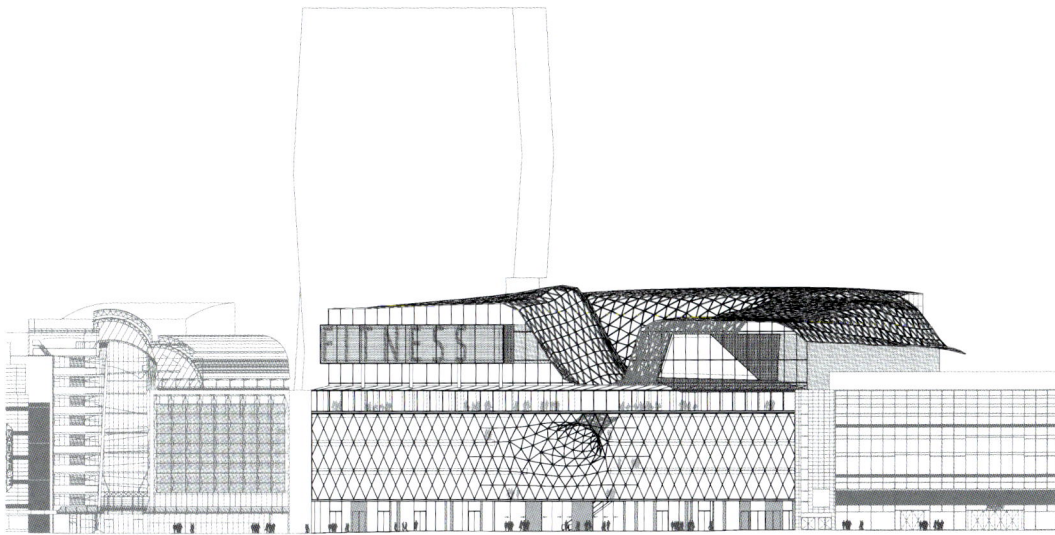

Ansicht Süd

Ansicht West

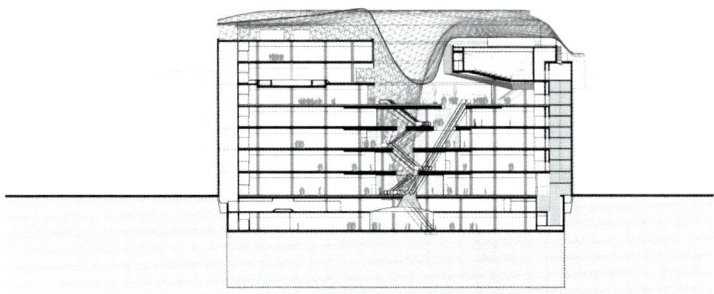

AA

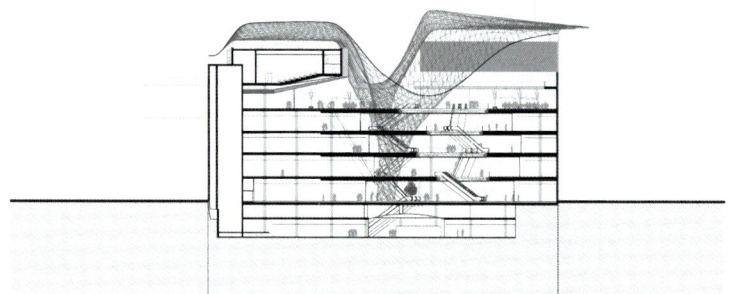

BB

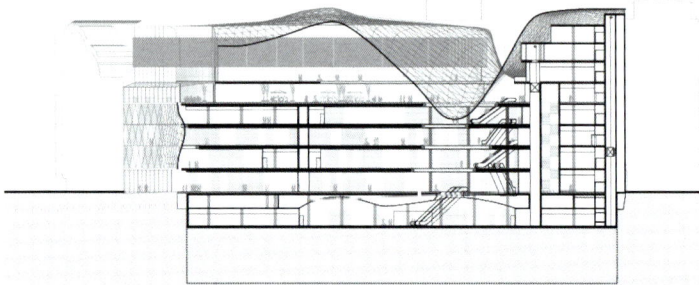

FF

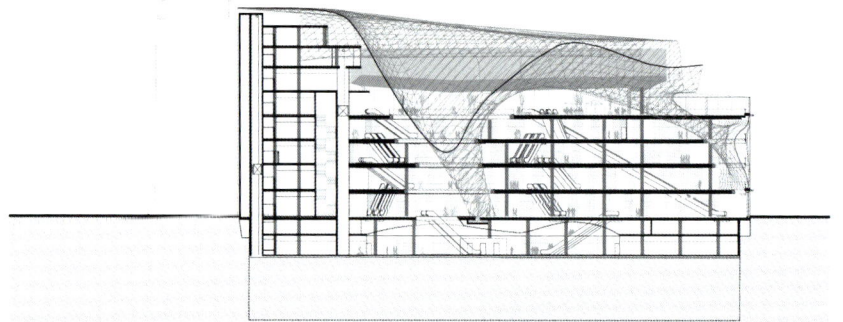

II

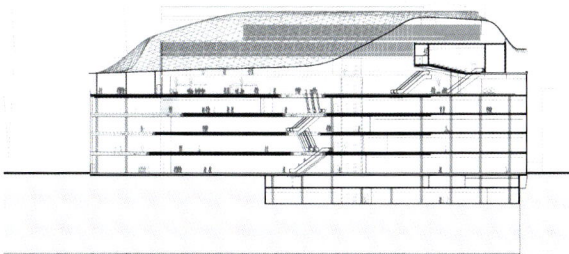

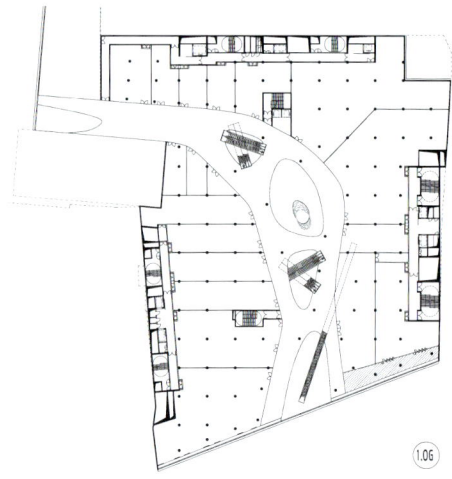
1.OG

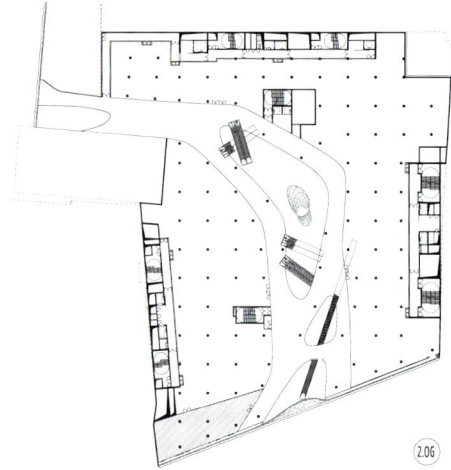
2.OG

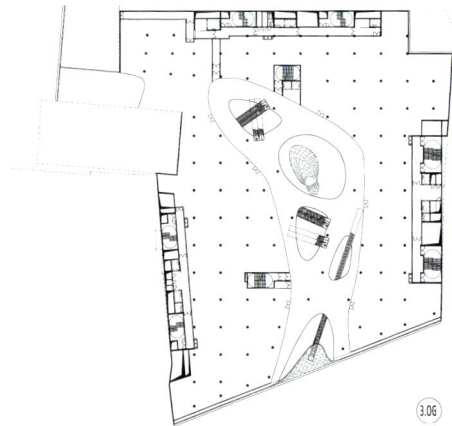
3.OG

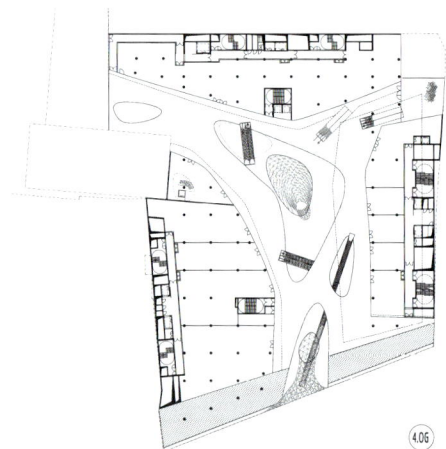
4.OG

5.0G

6.0G

7.0G

8.0G

ZG

1.UG

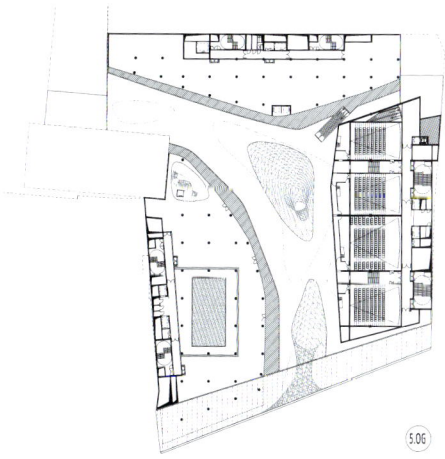

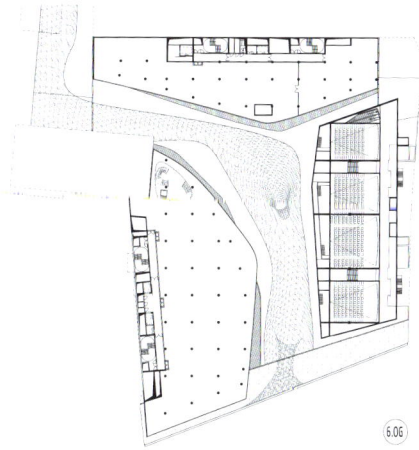

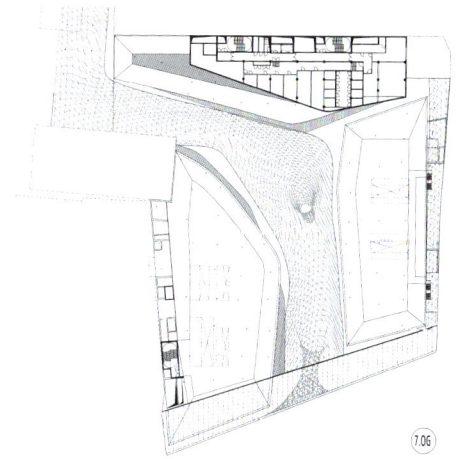

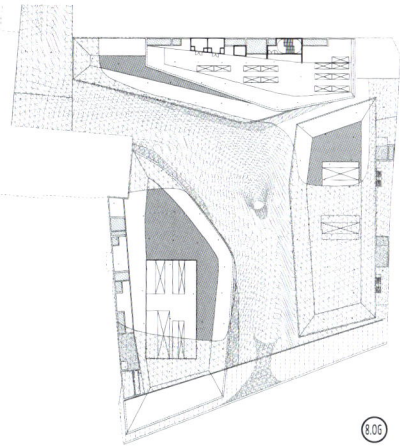

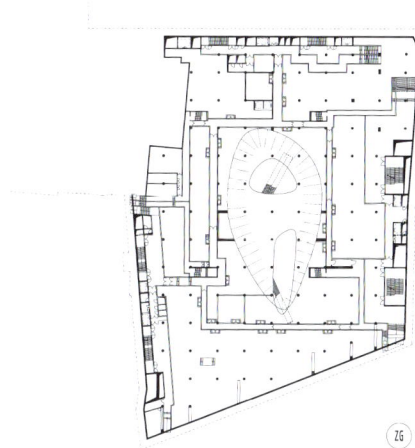

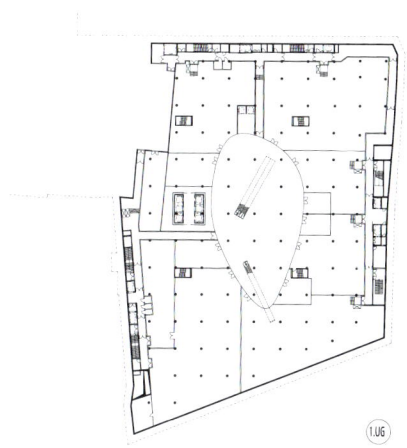

Frankfurt Hoch Vier购物中心的基本设计理念是建立一个垂直的城市，一条有垂直性这个新元素的城市购物街。大楼内部以及法兰克福城市景观的物理和视觉的结合把这个庞大的项目整合到法兰克福市中心的城市规模里。

弧线造型屋顶的主立面和有空洞的峡谷的连接是该项目的特征元素。

连接的加强构造了法兰克福主要购物街蔡尔街的重要入口。它吸引人们进入这个流畅的内部空间，它不仅连接着蔡尔街和Thurn und Taxis宫以及在一旁的"法兰克福评论报"区，也把底层和多功能零售娱乐中心连接起来。进入大楼更深处，从五楼下来的排放系统把投射进来的光带给了人们。五楼直接与底层的高速公路相连，成为第二个一楼。它的角色是一条拥有一个有吸引力的休闲计划的明亮的自然人文城市购物街。五楼功能区的分布以及五楼以上楼层和酒店及附近办公大楼的连接创造了一个有生气、有吸引力的区域。

Armani Ginza Tower, Tokyo

Architect: Doriana & Massimiliano Fuksas
Firm: Studio Fuksas
Location: Tokyo ginza central
Area: 7370m²
Photographer: Nakasa &Partners Inc, Tokyo

LED-illuminated bamboo stalks and leaves adorn the façade of the new 12-story Giorgio Armani flagship store in Tokyo's Ginza district.

The design element from the Emporio Armani retail areas—black stainless-steel panels with laser-cut slots backlit with fluorescent lamps—is carried through tothe common stair of the Tokyo subway stop that sits directly beneath the Armani store location.

Gold mesh, patterned with the bamboo leaf motif, is sandwiched between glass panels that line the stairs that lead from the ninth-floor restaurant up to the 11th-floor private bar called Privé.

Like a fine piece of fabric that Armani would use to create a garment, gold mesh sandwiched in between glass panels form the walls of a private dining area in the restaurant.

Case good merchandise in the Emporio Armani store areas is highlighted from a concealed slot system in the ceiling using CDMR111 lamps.

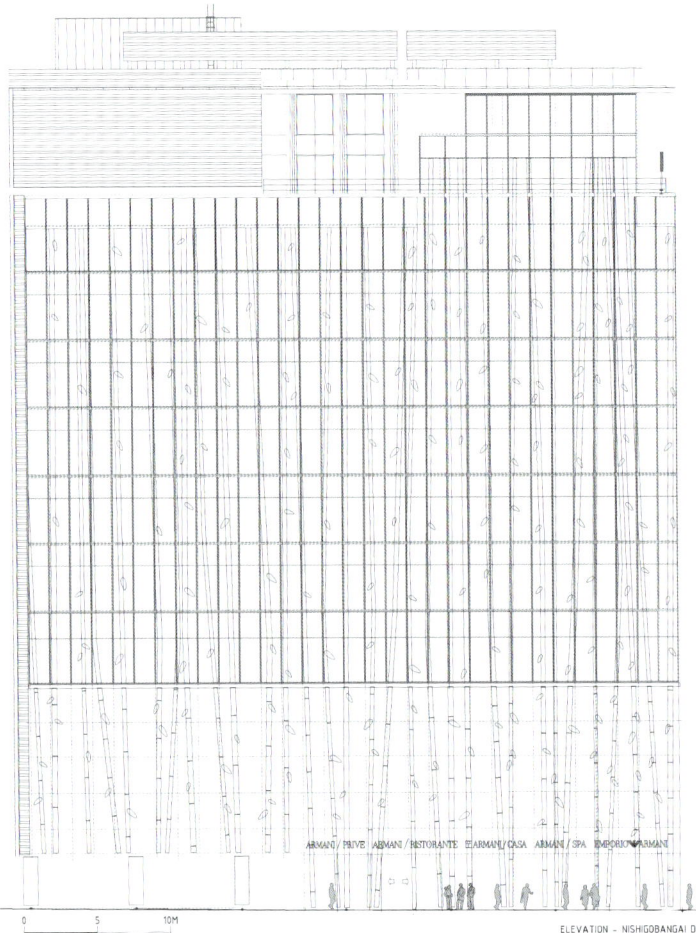

ARMANI / PRIVE | ARMANI / RISTORANTE E | ARMANI/CASA   ARMANI / SPA   EMPORIO ARMANI

0   5   10M

ELEVATION - NISHIGOBANGAI DI

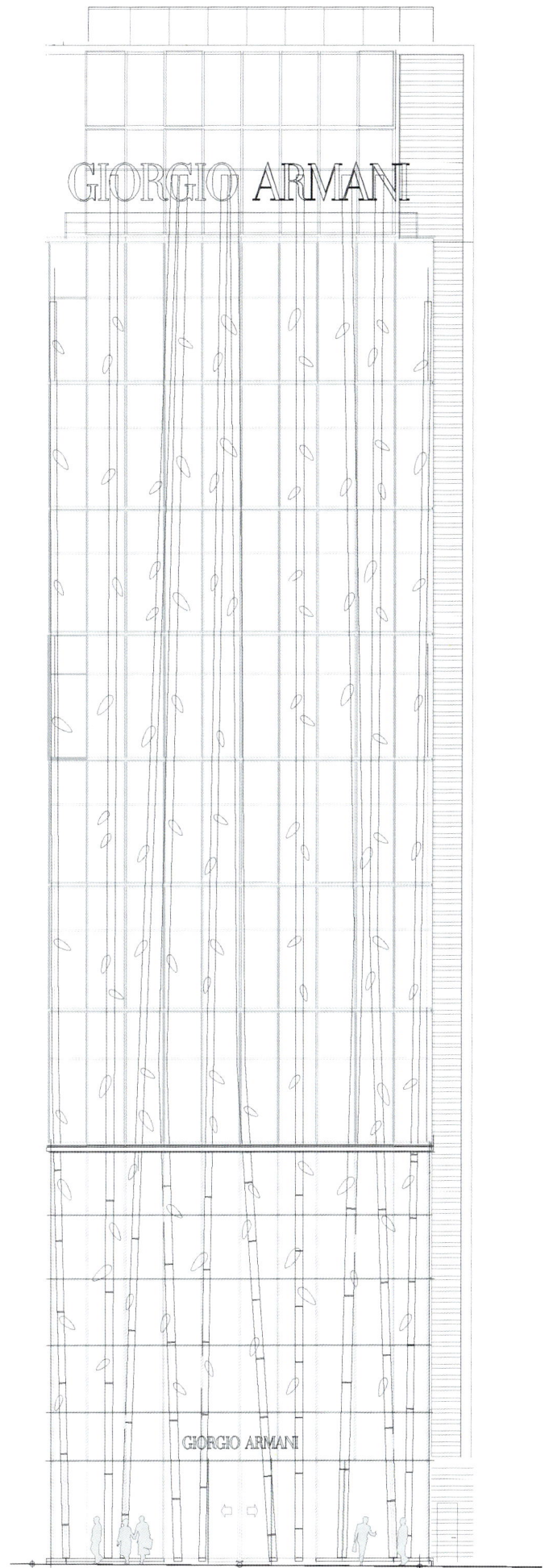

ELEVATION - HARUMI DORI

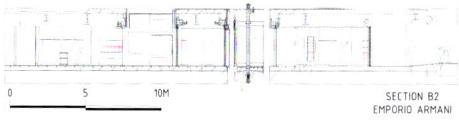

SECTION B2
EMPORIO ARMANI

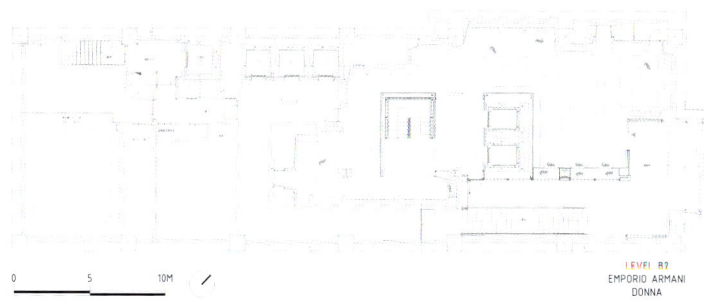

0   5   10M

LEVEL B2
EMPORIO ARMANI
DONNA

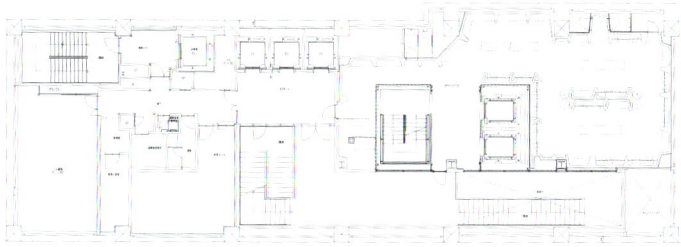

0   5   10M

LEVEL B1
EMPORIO ARMANI
-UOMO-

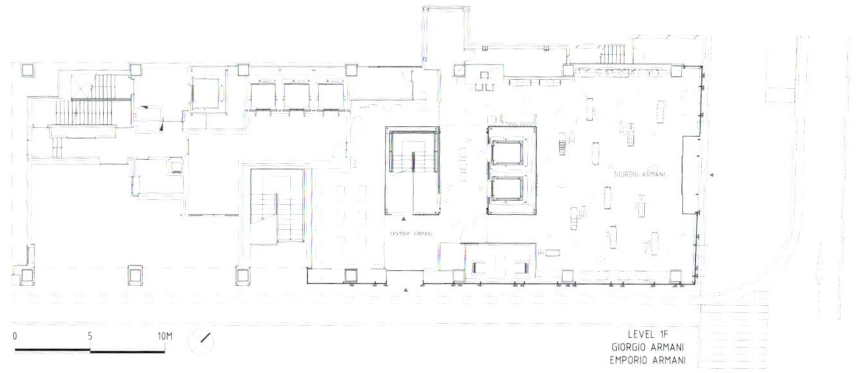

0   5   10M

LEVEL 1F
GIORGIO ARMANI
EMPORIO ARMANI

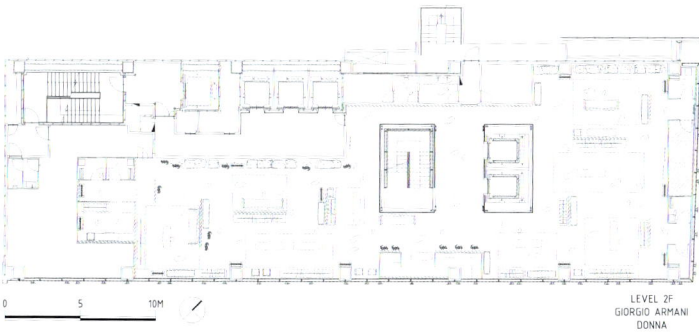

0   5   10M

LEVEL 2F
GIORGIO ARMANI
DONNA

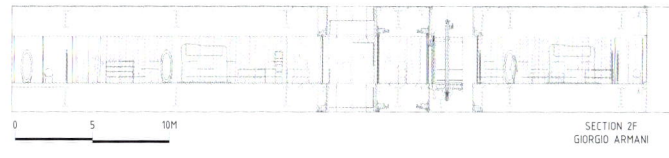

0   5   10M

SECTION 2F
GIORGIO ARMANI

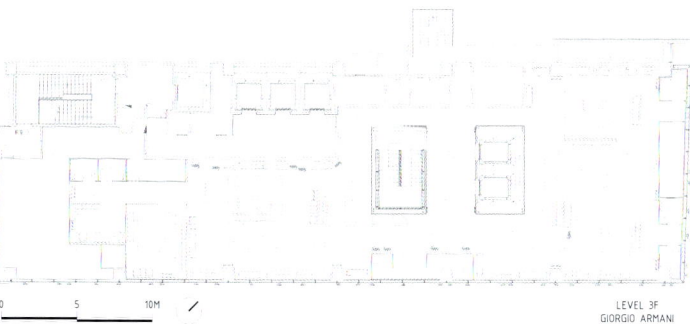

0   5   10M

LEVEL 3F
GIORGIO ARMANI
UOMO

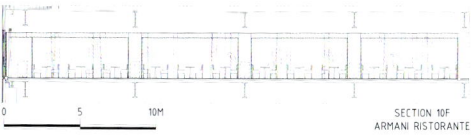

0   5   10M

SECTION 10F
ARMANI RISTORANTE

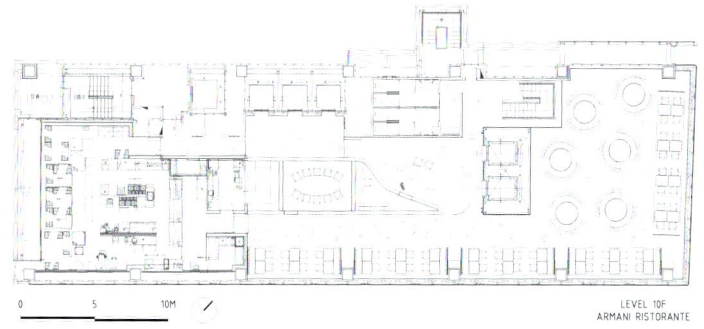

0   5   10M

LEVEL 10F
ARMANI RISTORANTE

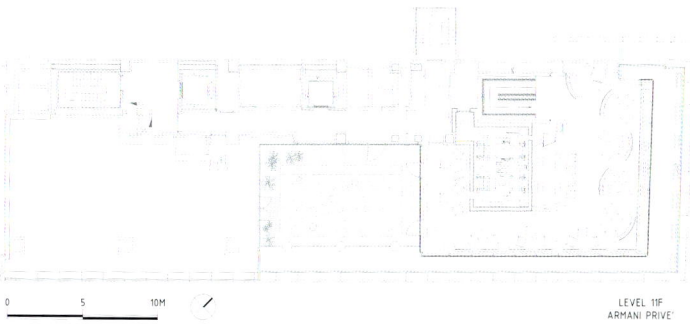

0   5   10M

LEVEL 11F
ARMANI PRIVE'

0   5   10M

SECTION A-A

SECTION B-B

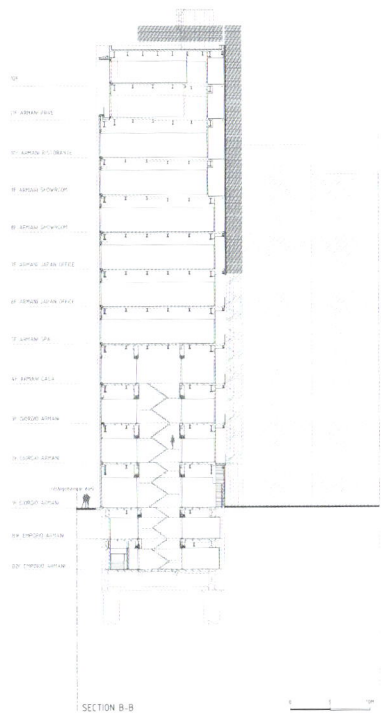

Giorgio Armani是位于日本银座的品牌旗舰店，其门面用采用竹竿和竹叶形状的发光二级管制成的灯装扮。

Emporio Armani零售区域的设计元素是背面带有荧光灯照明的黑色不锈钢板与激光切割槽，该元素取材于阿玛尼店铺正下方的东京地铁站的普通阶梯。

双层玻璃片夹着饰上竹叶的金丝网充当楼梯扶手，一直从9楼的餐馆延伸到11楼的Prive私人酒吧。

用双层玻璃夹着金丝网充当餐厅内私人就餐区域的墙壁，就如同Armani将一匹好布做成服装。

Emporio Armani天花板的隐槽内装着CDMR111灯，在灯光的映射下，存放区上的商品显得分外突出。

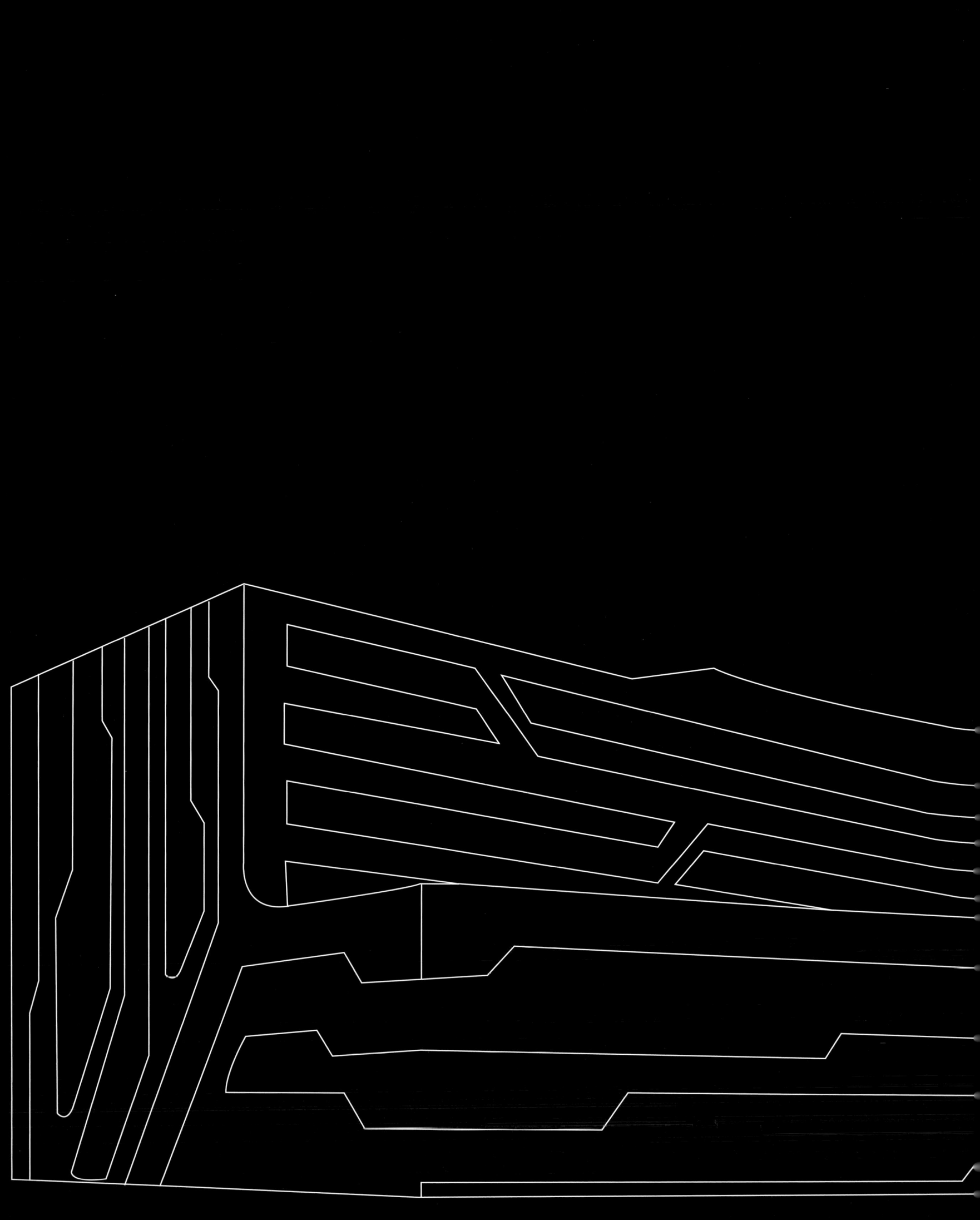

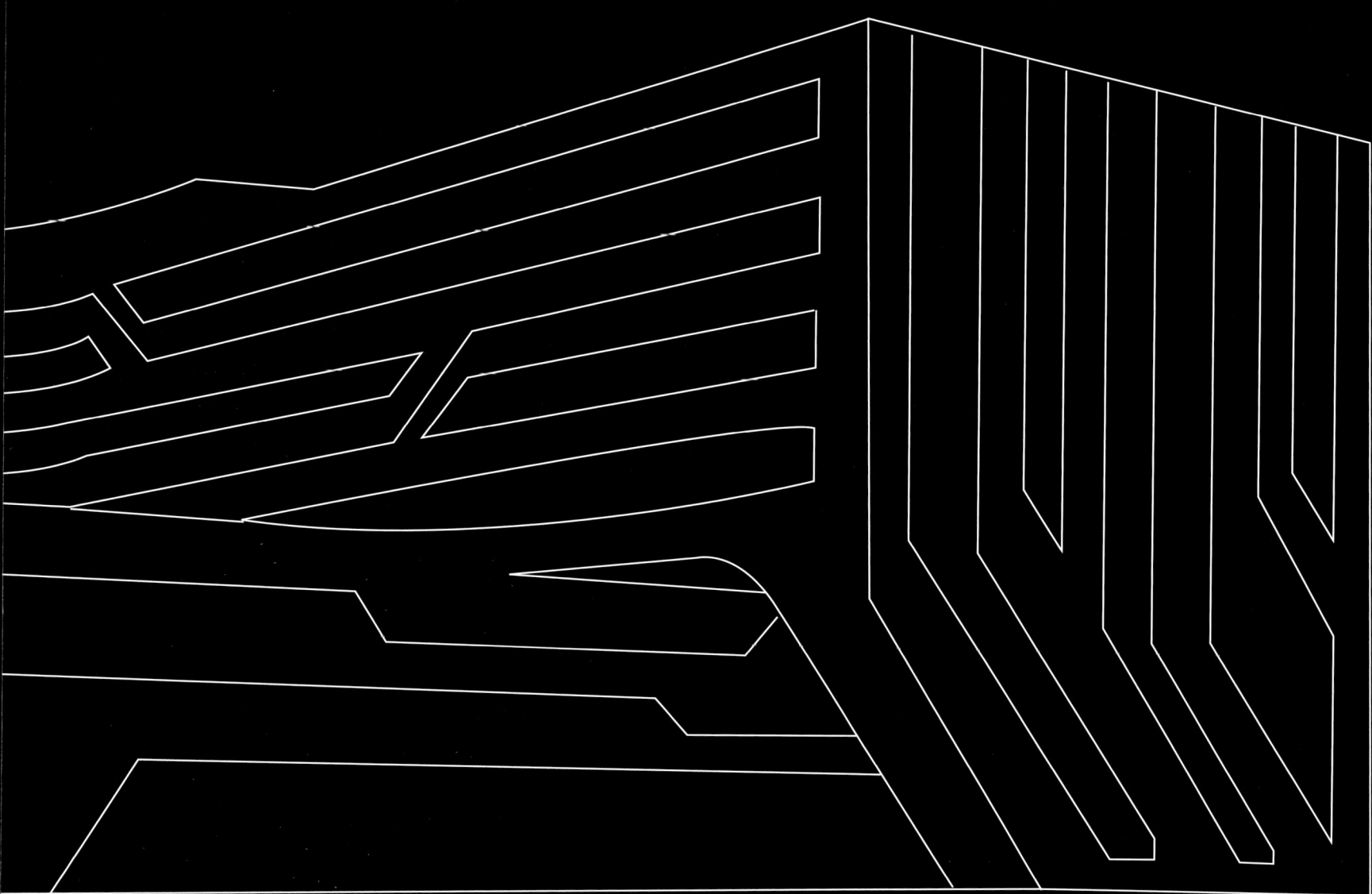

SOF Park Inn Hotel Complex

Architect: Juergen Mayer H., MarcusBlum, Jan-Christoph Stockebrand, Wojciech Witek, Magdalena Skoplak-Seweryn, Jakub Kaczmarczyk
Firm: J. MAYER H., GD&K Consulting Sp. z o.o., OVOTZ Design LAB
Location: Krakow, Poland
Photographer: Jakub Kaczmarczyk, Ovotz design Lab

SOF Hotel was designed by J. MAYER H. Architects is located in Krakow, Poland. Krakow, lying on the Vistula River in the south of the country with a population of about three-quarters of a million, is also an important center for culture and business in Poland with a bustling tourist trade heavily influenced by its historical architecture, such as Wawel Royal Castle and St Mary's Basilica dated from the 11th and 13th centuries.

Designed by J. MAYER H. Architects of Berlin with OVOTZ design Lab, the hotel's black and white exterior with its horizontal lines, clad of aluminium interspersed with black windows, is quite stunning. The sleek modern geometric lines are carried over into the interior of the hotel, particularly in the reception area where the massive reception desk is a focal point. 152 rooms and 10 Junior suites await their guests with all the usual add-ons now expected of any business and tourist hotel, such as a ballroom for conferences and a fitness room where guests can destress at the end of another busy day. From this central location, and lying close to the Jewish District of Kazimierz, the hotel undoubtedly offers a fabulous view over the city centre.

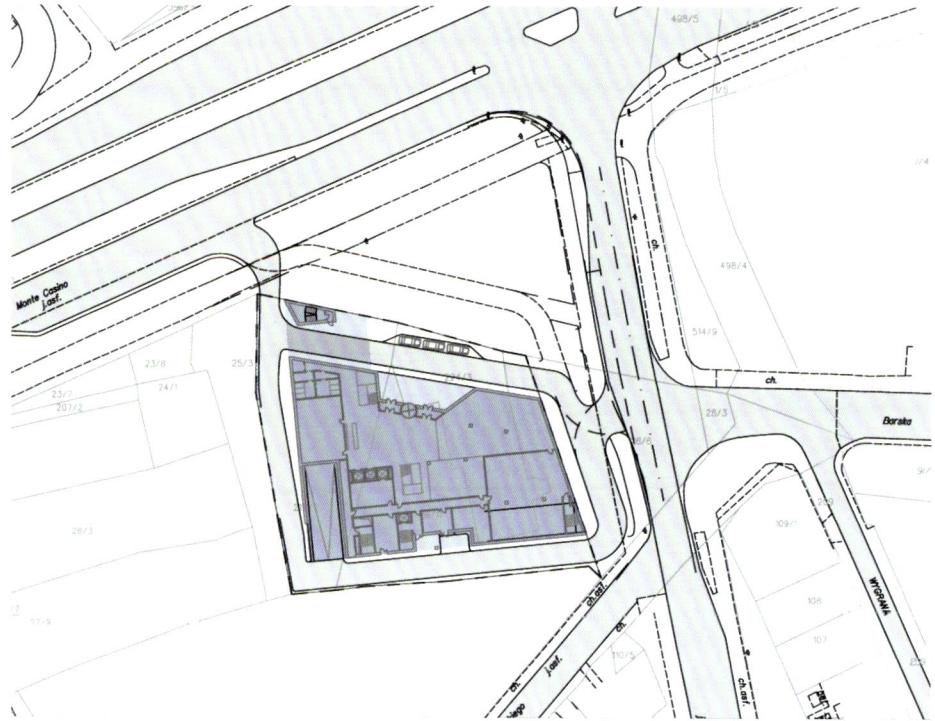

255 cm	
255 cm	
255 cm	
255 cm	
70 cm	
330 cm	
70 cm	
330 cm	
270 cm	

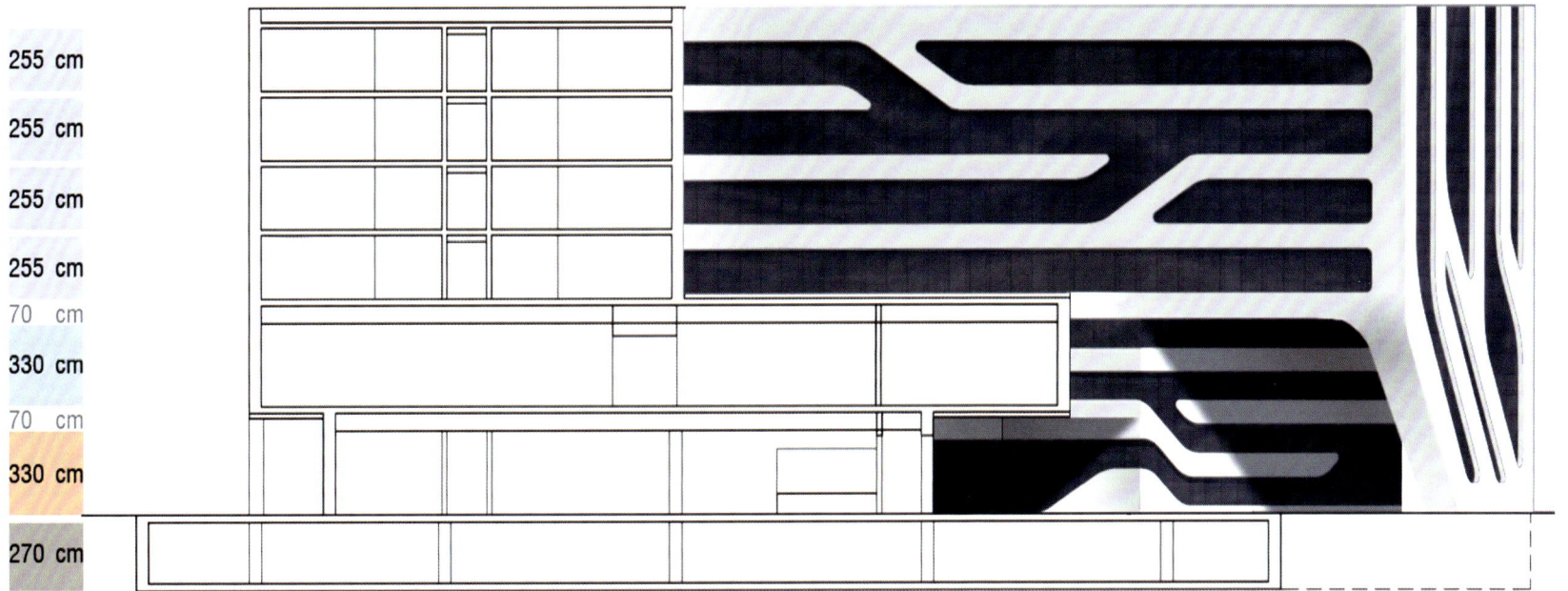

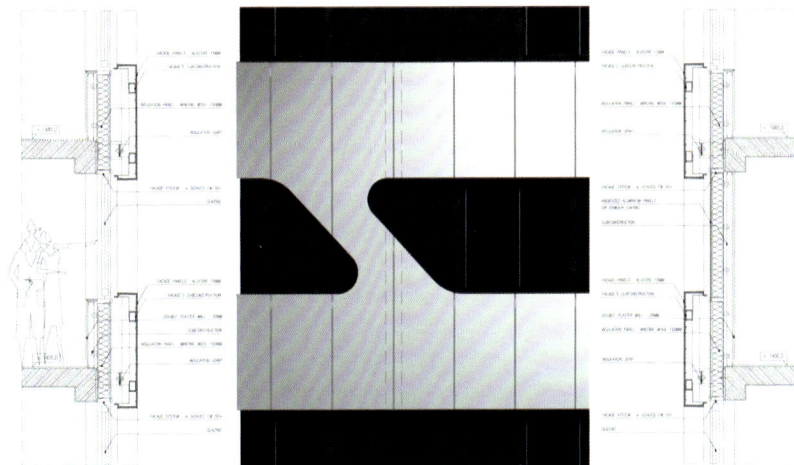

　　由柏林J. Mayer H.建筑事务所设计的SOF酒店，坐落在波兰的克拉科夫。克拉科夫位于波兰南部的维斯瓦河旁，它的人口总数占了全国总数的四分之三，它除了是波兰重要的文化和商业中心以外，还是一个深受历史建筑（例如11世纪的瓦维尔皇家城堡和13世纪的圣路易大教堂）影响的繁华的旅游城市。

　　这座建筑的设计师十分出色，是由柏林J. Mayer H.建筑事务所和OVOTZ设计实验室共同设计的。这座酒店的外墙用黑白色水平线条点缀着黑色窗户的铝制覆面。光滑的现代几何线条也同样延伸到了酒店内部，接待区的巨大台面是一个焦点。饭店有152间客房和10间套房，设置了舞厅、会议室、健身室等，忙碌了一天的顾客在这里能充分消除紧张的压力。该酒店靠近市中心和卡奇米日犹太区，无疑将成为市中心的一大景观。

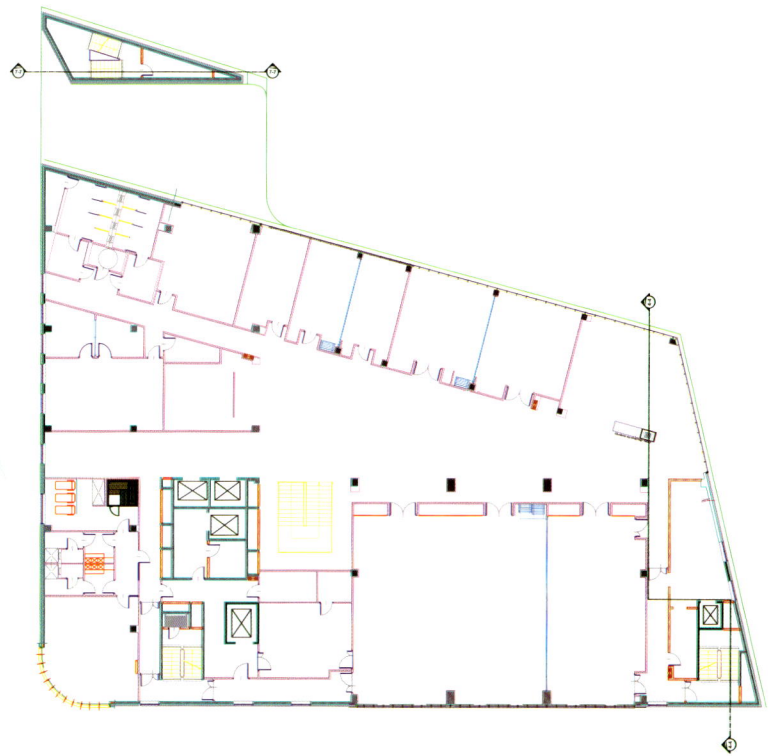

HOF Vitoria

Architect: Antonio Vaíllo i Daniel - Juan
L. Irigaray Huarte, David Eguinoa
Firm: Vaillo + Irigaray Arquitectos
Location: Vitoria
Photographer: José Manuel Cutillas

The building's location renders it a "door" or space between the frames of reference cities, and it maybe get its force due to its peculiar location, where the "landscape" has always been essentially flat: factories, warehouses, barracks...

This would give a single reading of the action, despite the different geometry, volume and use of the two buildings. It proposes the use of similar materials but with different configurations: aluminium lattice surrounds glazed facades. The lattice allows a double reading: the unification of the two volumes and the perception of the composition of each, depending on the angle of vision: from afar or from opposite the lattice disappears and inside the box and the glass shades are exposed; the leaned vision perspective offers a compact volume of "coarse grain", which outlines the pure geometry of each body.

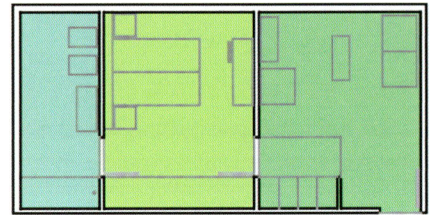

APARTAMENTO TIPO A

- ESTAR-COMEDOR
- DORMITORIO
- ASEO

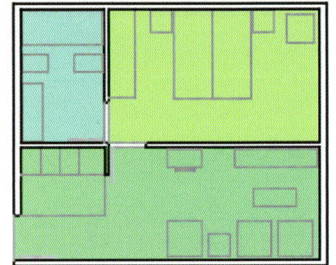

APARTAMENTO TIPO B

- ESTAR-COMEDOR S/NORMATIVA (12
- DORMITORIO
- ASEO

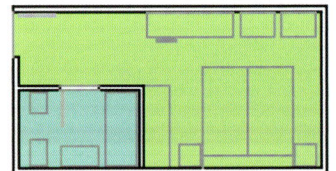

DORMITORIO TIPO

- DORMITORIO
- ASEO

HUECOS DE FACHADA
COMPOSICIÓN DE FACHADA

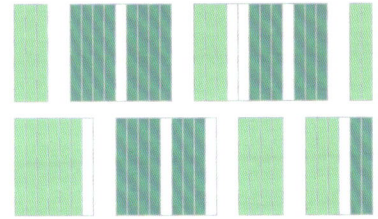

U-GLASS COLORES

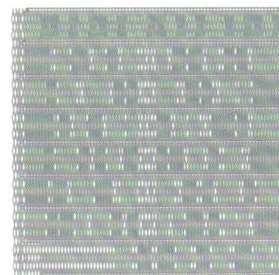

ALZADO SUR e:1/750

LEYENDA DE COLORES

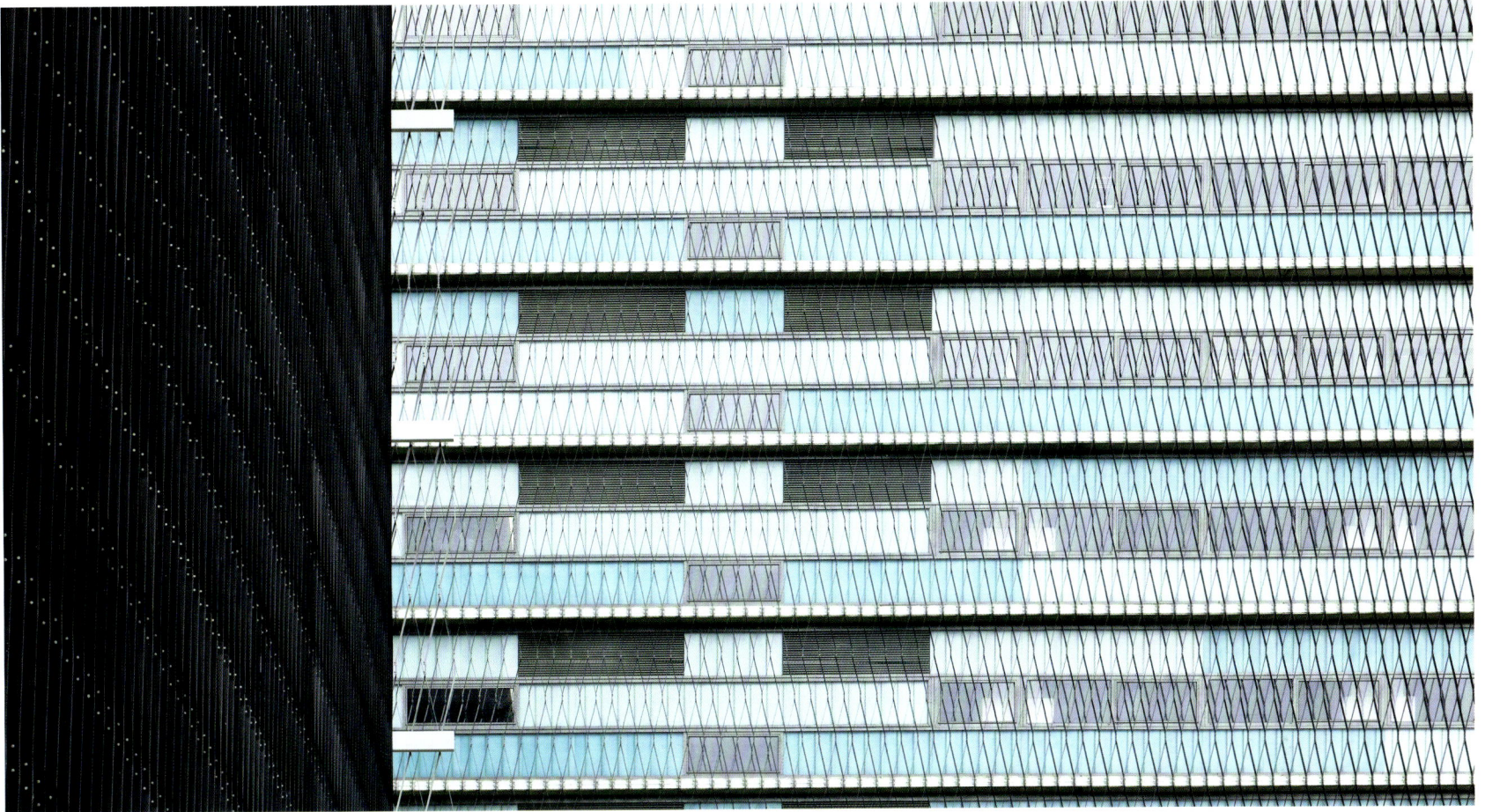

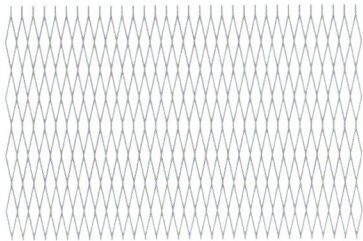

REJILLA ALUMINIO

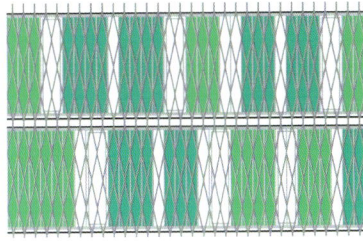

CONJUNTO FACHADA

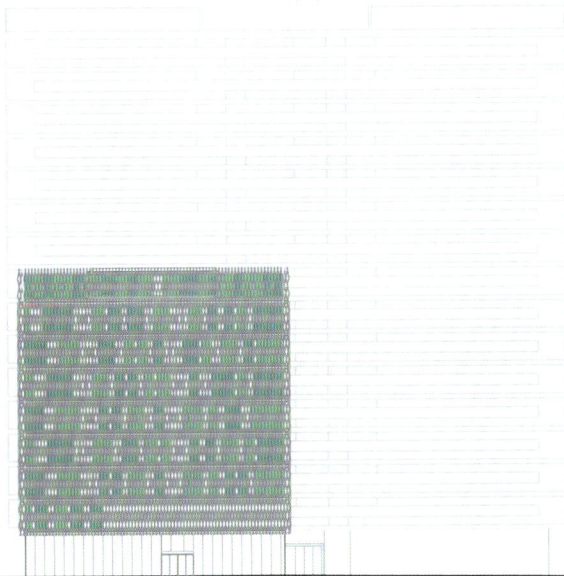

ALZADO OESTE e:1/750

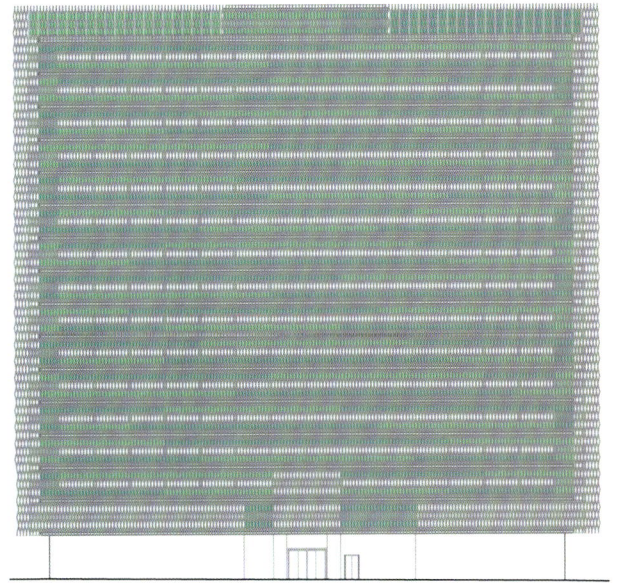

ALZADO ESTE e:1/750

25m

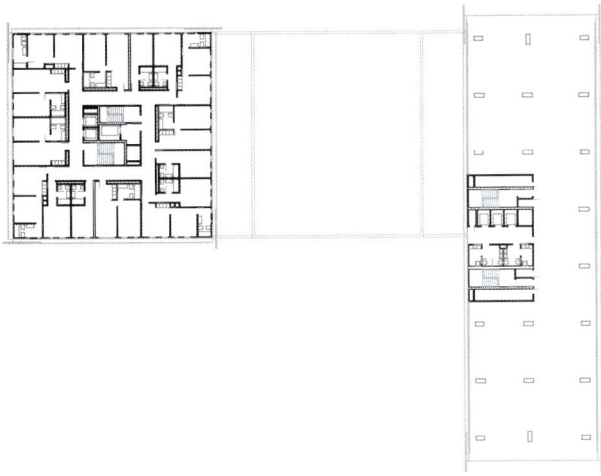

PLANTA TIPO e:1/500

0 2   10   25m

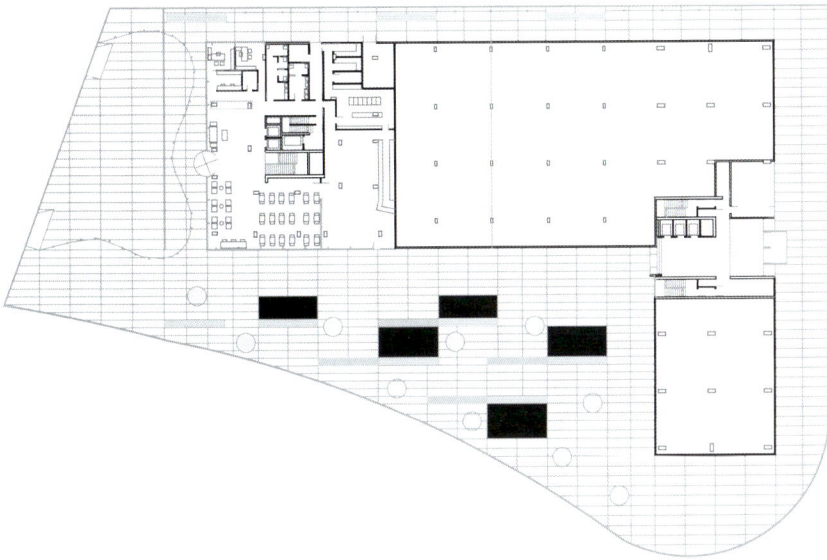

PLANTA BAJA Y URBANIZACIÓN e:1/500

0 2   10   25m

　　该建筑的位置使它成为一道"门"或是渲染周围城市框架之间的空间，或许它可以因为它独特的地理位置得到加强，但周围的景观本质上是单调的：工厂、仓库、营房……

　　尽管有不同的几何形状、体积并且两座建筑物的作用也不同，但都会有同一种解读。设计提出使用相似的材料，但使用不同的配置：铝制晶格包围着光滑的外立面。晶格给出双重解读：两种体积的统一和每种构成的感觉，这些都取决于这样的视角：从远处或对面看去，晶格逐渐消失，盒子的内部和玻璃阴影显露出来。倾斜的视觉角度提供了一个小体积的"粗晶粒"，它诠释了每个主体的纯几何形状。

WISA Wooden Design Hotel

Architect: Pieta Linda
Location: Helsinki

The WISA Wooden Design Hotel is an outstanding example of wood's versatility. The first thing that catches the eyes is the atrium courtyard in the center. The long curved pine boards half-covering the patio form a trellis that titillates the imagination. The trellis protects against the wind and filters the long rays of the Nordic sun into the courtyard. The charming interplay of light and shade can be observed throughout the day, as the bright morning light gradually turns into the red glow of the Northern midnight sun. The striking wood architecture makes its way to the living quarters. The pine floor is complemented by the light ethereal beauty of Nordic birch. The walls and the ceiling are paneled with beautiful and durable birch plywood. Light comes in at both ends of the living quarters through floor-to-ceiling windows that brighten the interior. From the bedroom you can admire the morning sun and the sea, while the view from the lounging area opens onto the city and the evening sun.

WISA Wooden设计酒店是个很出色的木制建筑的范例，成功地展现出木质的多变。第一眼吸引人的就是那位于中心的中庭，那又弯曲又长的松木条，半遮掩天井交叉又成格子状，能激发人们的想象力。这个棚架能逆风保护建筑并对长期进入庭院的北欧阳光辐射进行过滤。在这里，一整天都能观察到迷人的光影作用，从早晨的阳光逐渐演变成北部午夜灼热的阳光。这个引人注目的建筑成功地融入到周围的生活区里。北欧桦木的空灵之美是对松木地板的补充。墙壁和天花板是由持久耐用又美观的桦木夹板镶嵌的。光线能从生活区两边尽头的落地窗渗入，使室内变得明亮无比。在房间里能欣赏到早晨的阳光和大海，还能看见城市的休闲区和傍晚落日的余辉。

Akmani Hotel

Architect: TWS & Partners
Location: Jakarta, Indonesia
District: Pusat
Site Area: 994m²
Bldg. Area: 1478m²
Gross Floor Area: 10.467m²
Blag. Coverage Ratio: 67.5%
Gross Floor Ratio: 70.82%
Photographer: Fernando

The project is briefed as a boutique – business hotel located in a main busy road in central of Jakarta, which is called Wahid Hasyim street.

Mainly known as a tourism and leisure, commercial stripe, this area contains many street side coffee shops and small hotels, which cater many local or foreigner tourists.

The area surrounding is allocated for the commercial usage which results in a high building floor ratio, because it has a direct access to the central business district of Jakarta, called Thamrin area.

The site itself is relatively small parcel of land, for commercial usage, which around 1500 square meters. Sandwiched between two commercial usage, a Spain embassy and office building, the site can be seen and approached easily from Thamrin area.

Taken this commercial consideration, the client wants to build a custom – self operated business hotel, which has a vision to deliver an unique and customize product to the market.

In order to deliver the best product to win over the market between surrounding competitors, the client held informal competition by end of 2007 to look for concept that suit for the project.

TWS & Partners succeed to convince client and got the project after few months process of presentation.

该项目的主旨是构建一所带小商品零售的商务宾馆，该宾馆位于雅加达一条繁华街道的中心，这条街被称为Wahid Hasyim街。

这条街以旅游休闲和商品零售而著称，有咖啡店、小旅馆，非常适合当地和外地游客游览、观光、居住。

建筑周边是具有商业用途的高大建筑物，直通雅加达商务区中心——Thamrin区。

作为商业用途，该建筑本身占地相对小一点，大约1500平方米。该建筑被夹在西班牙大使馆和一座办公大楼之间，在Thamrin区很容易看到并到达这栋大楼。

出于商业考量，客户希望建立一个顾客自助式的商业酒店，并希望可以为市场提供一种独特的定制产品。

为了提供最好的产品以便在和周边的竞争对手的较量中赢得市场，客户于2007年年底举行了非正式的竞标，希望找到最适合这个项目的设计理念。

经过几个月的努力，TWS & Partners终于成功地说服了客户，拿到了这个项目。

San Ranieri Hotel

Architect: Simone Micheli
Firm: Simone Micheli Architectural Hero
Location: Pisa, Italy
Area: 5387m²

The building, whose externals are colored by an intense sea blue color and are almost totally covered with ethereal curtain of transparent/ translucid glass, becomes animated with the sunlight, openly manifesting itself in its completeness during the day and becoming of iridescent and impalpable tone's shades and colors during the night.

From the ceiling allowing the guest to redefine always the space, and to the sinuous continuity of the wall shaped with light white varnish plaster that divides technical function from the common ones and it transposes the concept of "space box" playing with light and shadow.

The rooms are a further evolution of this game between contrasts that stimulate fantasy and creative spirit; the standard rooms, dominated by the material of the pavement and of the light grey ceiling in cement, are typified by black varnished wooden furniture, contrasting between light and dark and with wall mirrors that becomes the principal spot lighting in the space. The color is the domineering theme also in the polychrome bathrooms of the rooms where there is no space for monotony and new atmospheres permeate the spirit. It's after this overwhelming whirlwind of emotions that the spirit of every guest will be inebriated by this charming and amazing tridimensional dream。

PIANTA PIANO TERRA
1 HALL
2 RECEPTION
3 DIREZIONE
4 AMERICAN BAR
5 RISTORANTE
6 SALA PRIVATA
7 CUCINA
8 LOCALE TECNICO
9 CAMERA DOPPIA

PIANTA PIANO TERRAZZA
10 CAMERA TIPO-B SUITE
11 SALA POLIVALENTE
12 TERRAZZA
13 LOCALE TECNICO

PIANTA PIANO TIPO
8 LOCALE TECNICO
9 CAMERA DOPPIA

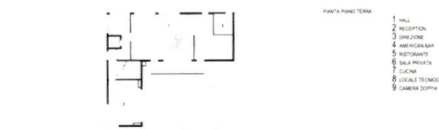

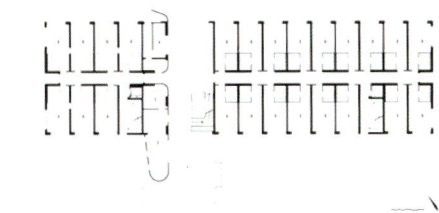

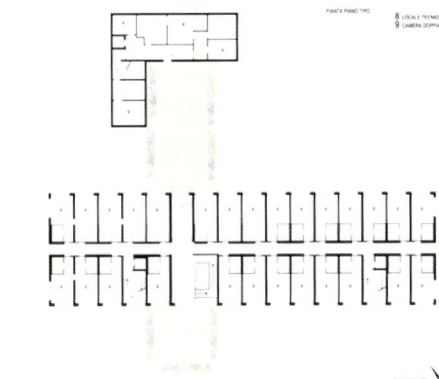

　　这座建筑拥有半透明的海蓝色玻璃外立面和通透的清玻璃屋顶。白天，沐浴在日光中的酒店毫无保留地向人们展示它的全部；入夜，变化的灯光从建筑内部照射出来，那可望而不可及的灯光把建筑装扮成一件迷人的艺术品。

　　从天花板延伸至地面的墙面不规则的弯曲起伏是用白色亮光漆石膏构成的，这个分隔墙与其他普通的工艺不同，再加上运用阴影和光亮的渲染效果，彻底颠覆了墙面传统划分空间的作用。

　　酒店的房间是激励的幻想和创新精神之间的对比游戏的进一步演化。标准客房的风格通过路曲和灰色的水泥天花板的材料来体现，以黑色的木质清漆家具作为典型的特点，墙壁安装了重要的聚光灯，形成强烈的对比感。以空间颜色为流行主题，这也体现在多彩的浴室，那里没有千篇一律的空间，创新精神的崭新气息渗透到了每一个角落。

CitizenM Hotel Amsterdam City Center

Architect: Rob Wagemans, Erikjan Vermeulen, Jeroen Vester, Sander Vredeveld, Matthijs Hombergen
Firm: Concrete Architectural Associates
Location: Amsterdam, Netherlands
Area: 6000m²
Photographer: Ewout Huibers, www.ewout.tv, Concrete Architectural Associates

CitizenM is a new Dutch hotel group. The concept of the hotel is to cut out all hidden costs and remove all unnecessary items, in order to provide its guests a luxury feel for a budget price.

The hotel does not have a service desk but a table with six self-check-in terminals. The total building is lowered because of the development regulations in the environment. The building is a black metal box, dominated by the pushed out big glass windows of the rooms. The various depths of the aluminum frames and the angled glass give an individual twist to the rigid façade. The big glass windows on the ground floor are placed inwards the building, creating a natural transfer between inside and outside and showing the living rooms and lobby space to the street.

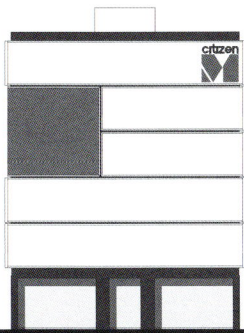
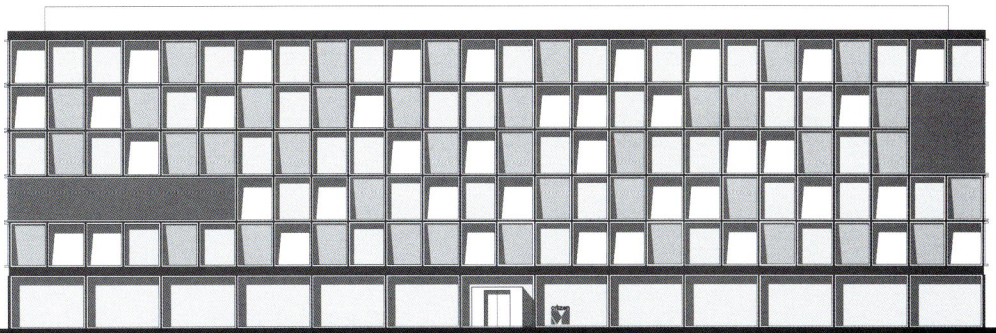

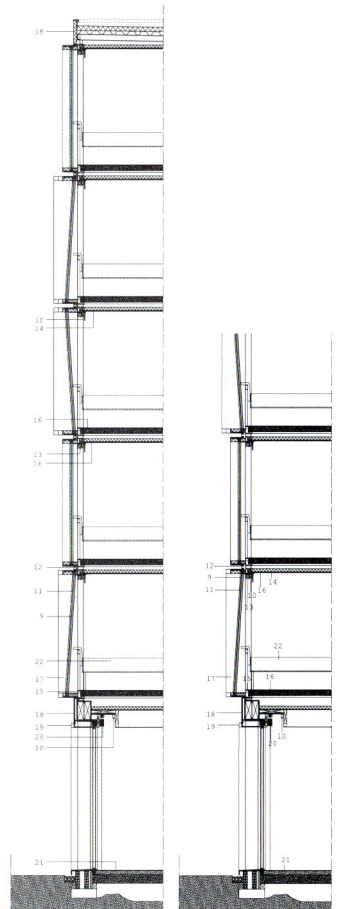

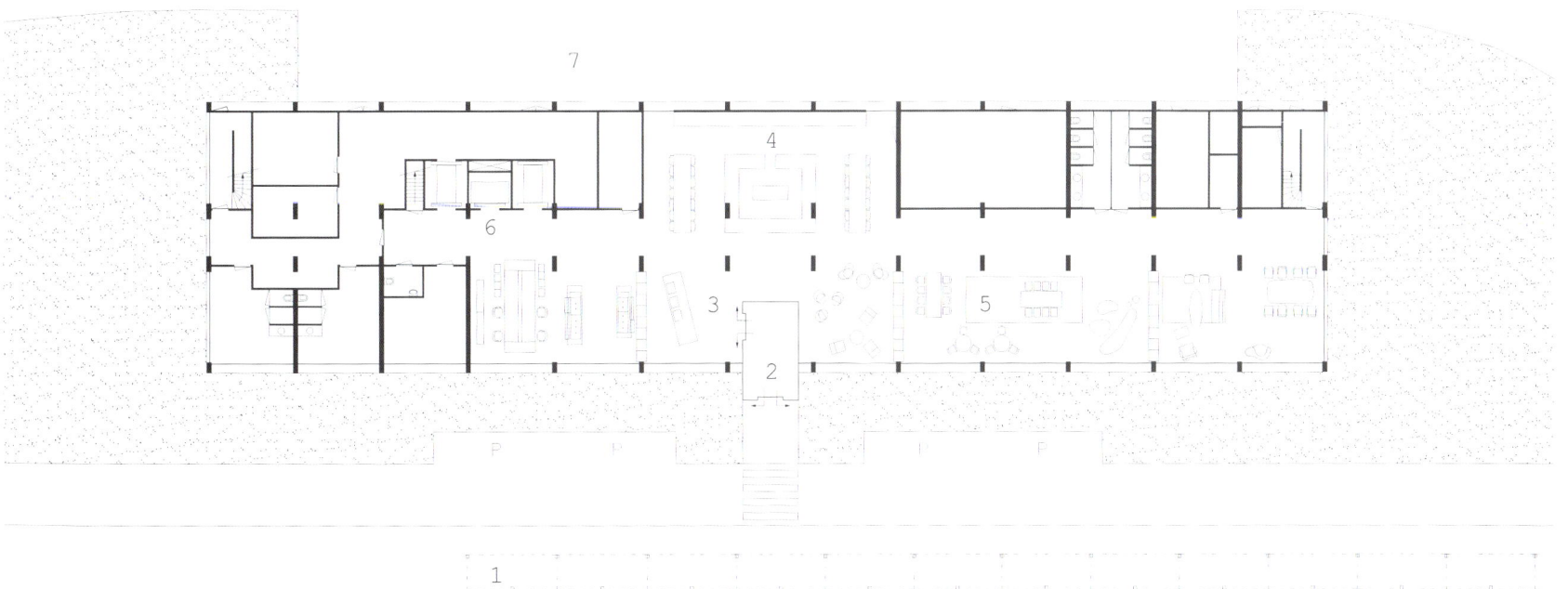

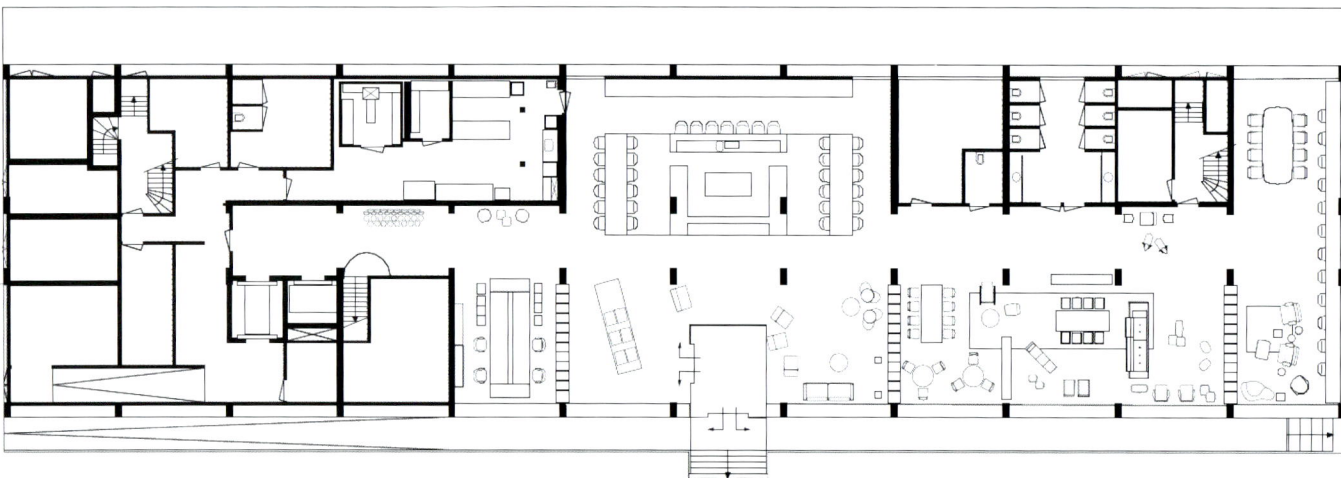

beethovenstraat

prinses irenestraat

1   covered course
2   entrance
3   check-inn
4   canteen M
5   livingrooms
6   elevators
7   supply
8   artwork
9   screenline inside glasspanel
10 luminaire
11 double glazing with -5, -2, 0, 2, 5
    degree inclination
12 black rubber cover fillet
13 black-out cutrain
14 gypsum fibre board
15 steelconstruction module
16 zebrano floor/ceiling
17 aluminium window frame, silver/
    metallic; depth: 250, 350, 450 mm
18 black coated composition
19 aluminium window frame
20 window blind
21 flooring: bamboo parquet/floating
    floor/ belgian bluestone

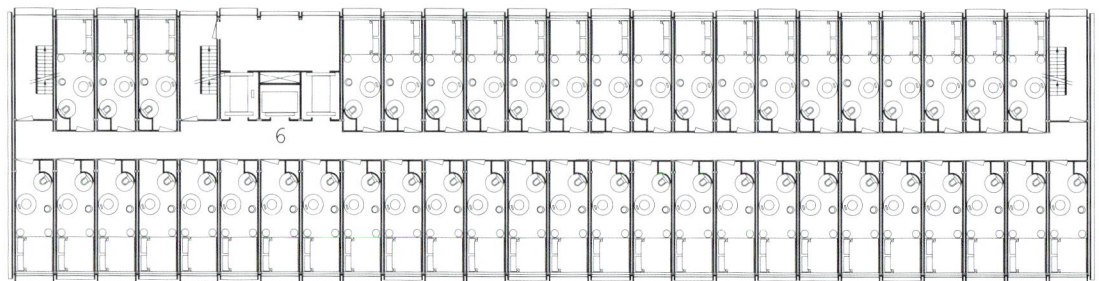

CitizenM是一个新生的荷兰酒店集团。该酒店的经营理念是免除所有的隐性消费，废去所有多余项目，让顾客享受平价的奢华服务。

酒店内没有服务台，但设有六个自助设备终端。设计师故意把酒店大楼设计的低一些是为了与周围环境相协调。酒店大楼看起来像一个黑色的金属盒子，房间外凸出来的大落地窗显得十分醒目。不同厚度的铝框和多角的玻璃给这个原本坚硬的建筑外表带来了灵动感。底层的落地窗内嵌在建筑立面，创造出了一种错落有致的建筑美感，也向街道上的人们展示了酒店的客厅和走道。

Mine Hotel Boutique

Architect: Adriana Sternberg & Mariano Kohen
Firm: Sternberg Kohen Arquitectos
Location: 4770 Gorriti – Buenos Aires City
Area: 932m²

FACHADA FRENTE

Mine hotel boutique has 20 rooms, in which pure expressions prevail. Net shapes, wood, rattan and whites with sparkles of colors that refer oriental.

The project's lines are based on two blocks of rooms, one on the front, and the other one looking toward the back garden, joined by vertical and horizontal bridge-like circulations, providing views to a central patio.

On the ground floor, a sequence of interior and exterior spaces separated by large glazed panels organize the different public functions from the main access, passing on to the lobby, the central patio, then the living room and the bar/restaurant and, at the end, the gallery, the deck and the back garden with a pool.

This allows to see all the spaces together, from the street to the end of the site, causing the guests a feeling of not knowing if they are inside or outside.

Mine hotel is located in the heart of Palermo Soho, shelter of the lovers of style and restless spirits.

Center of the new Argentina design, the Soho of Buenos Aires is surrounded by art galleries, clothes stores and design stores, cafes, bars and restos.

Plastic artists, designers, writers, poets, filmmakers, publicists and musicians come together in this center where history and modernity are merge.

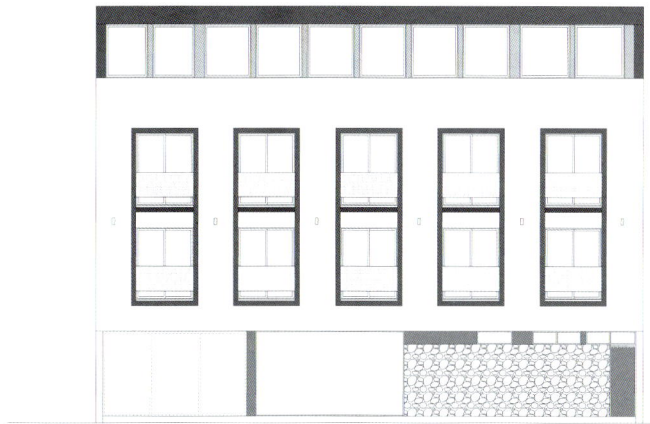

FACHADA CONTRAFRENTE

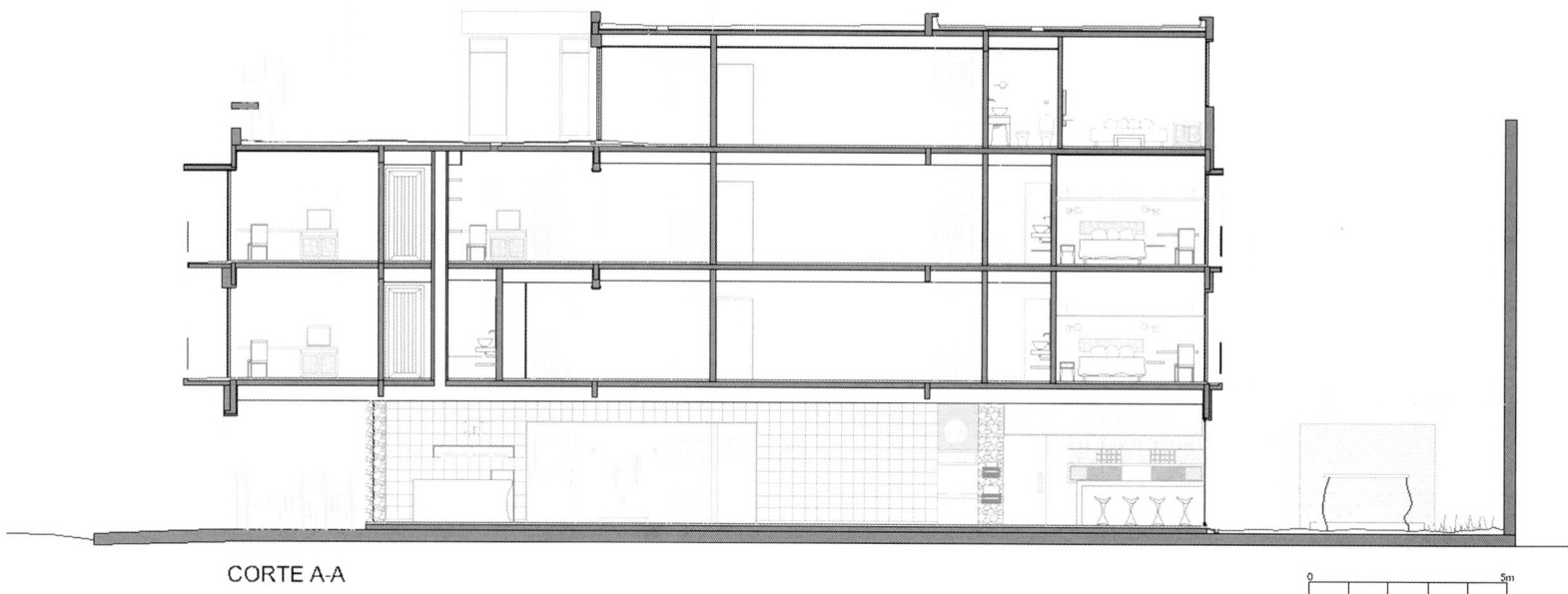

CORTE A-A

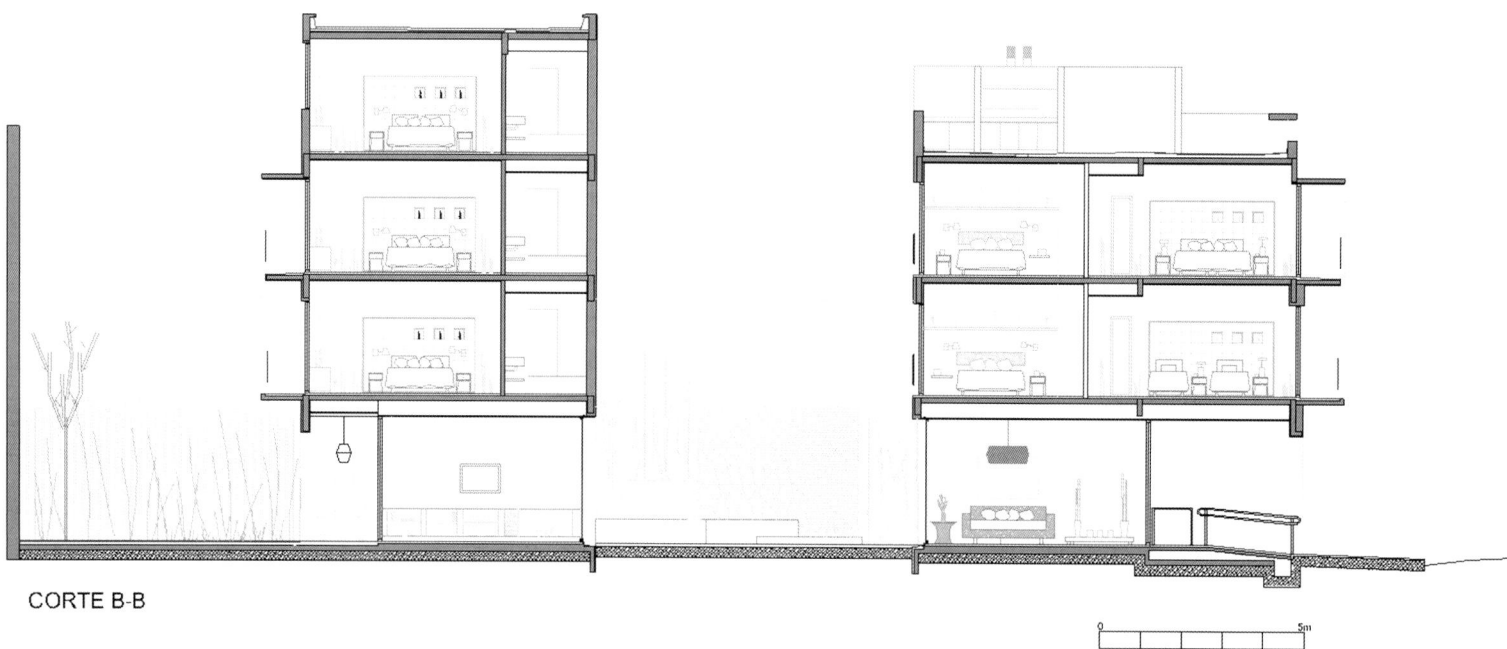

CORTE B-B

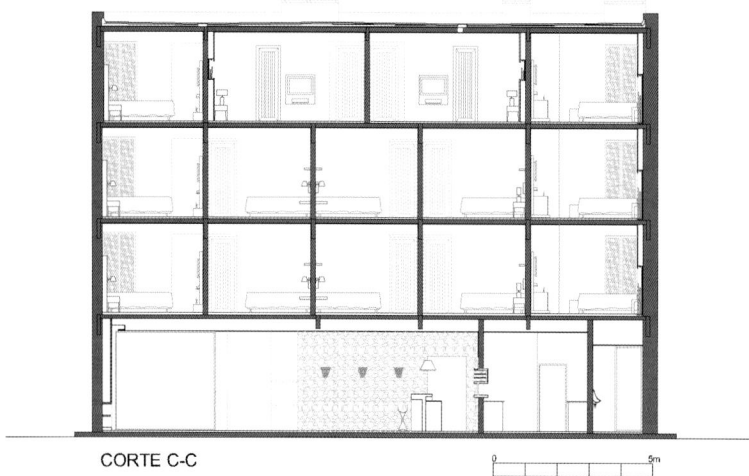

CORTE C-C

该精品酒店拥有20间客房，处处表现了流行元素。网格形状、木材、藤制品和白色与火花的颜色代表着东方。

该项目的主线是基于两个区域的客房，一个在前面，另外一个面向后花园，加入了纵向和横向桥式环道，在这里，可以看到中央露台的美景。

在底层，一系列内部和外部空间被大块的玻璃板分隔开来，以便从主入口进入后可以形成通往各功能区的通道，这样便可进入大厅、中央露台，然后是客厅和酒吧/餐厅，最后到达画廊、木制平台，以及带有游泳池的后花园。

这种设计使得人们可以看到建筑的各个空间，可以站在街头观赏到酒店的内景，这也使得客人们有一种不知身处室内还是室外的感觉。

该酒店位于巴勒莫苏荷中心，为那些崇尚个性和精神得不到休息的人提供休憩地。

阿根廷新的设计中心——布宜诺斯艾利斯苏荷四周围绕着画廊、服装商店和设计商店、咖啡馆、酒吧和餐厅。

塑造艺术家、设计师、作家、诗人、电影制作人、宣传人员和音乐家聚集在这个中心，体现出历史与现代的融合。

PLANTA BAJA

1. ACCESO - 2. LOBBY - 3. OFICINA - 4. TOILETTE - 5. PATIO -
6. CIBER - 7. LIVING - 8. RESTO - BAR - 9. COCINA - 10. AREA
PERSONAL - 11. JARDIN - 12. PISCINA

GORRITI

PLANTA PRIMER PISO

1. HABITACION SMALL - 2. HABITACION MEDIUM -
3. HABITACION LARGE - 4. BAÑO - 5. BAÑO CON JACUZZI - 5.
HABITACION DISCAPACITADOS - 7. BAÑO DISCAPACITADOS

0 _____ 5m

PLANTA SEGUNDO PISO

1. HABITACION SMALL - 2. HABITACION MEDIUM
3. HABITACION LARGE - 4. BAÑO - 5. BAÑO CON JACUZZI

0 _____ 5m

PLANTA TERCER PISO

1. HABITACION LARGE PLUS - 2. HABITACION MEDIUM
3. HABITACION LARGE - 5. BAÑO CON JACUZZI - 6. SALA DE
MAQUINAS - 7. LAVADERO - 8. PARRILLA

0 _____ 5m

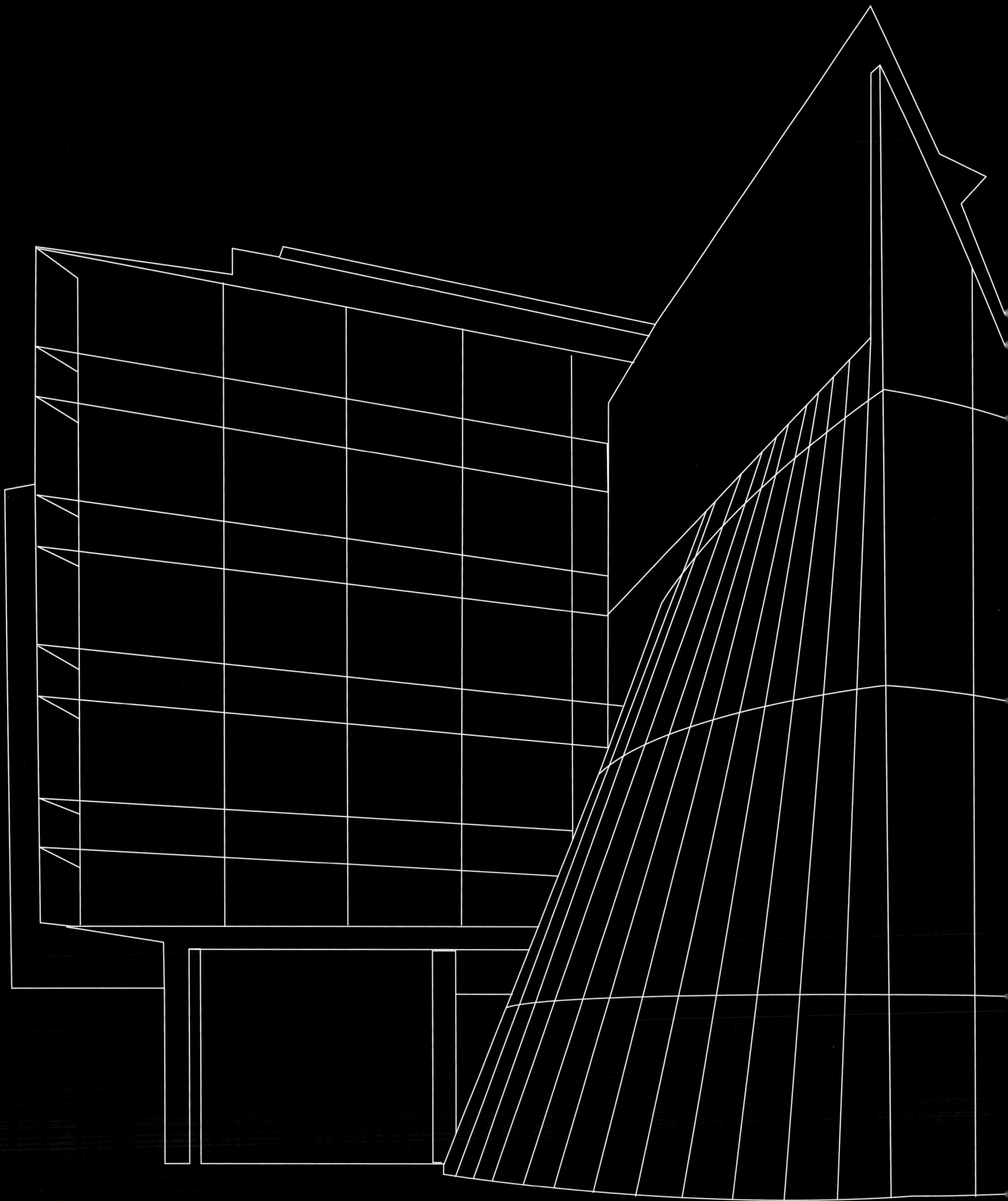

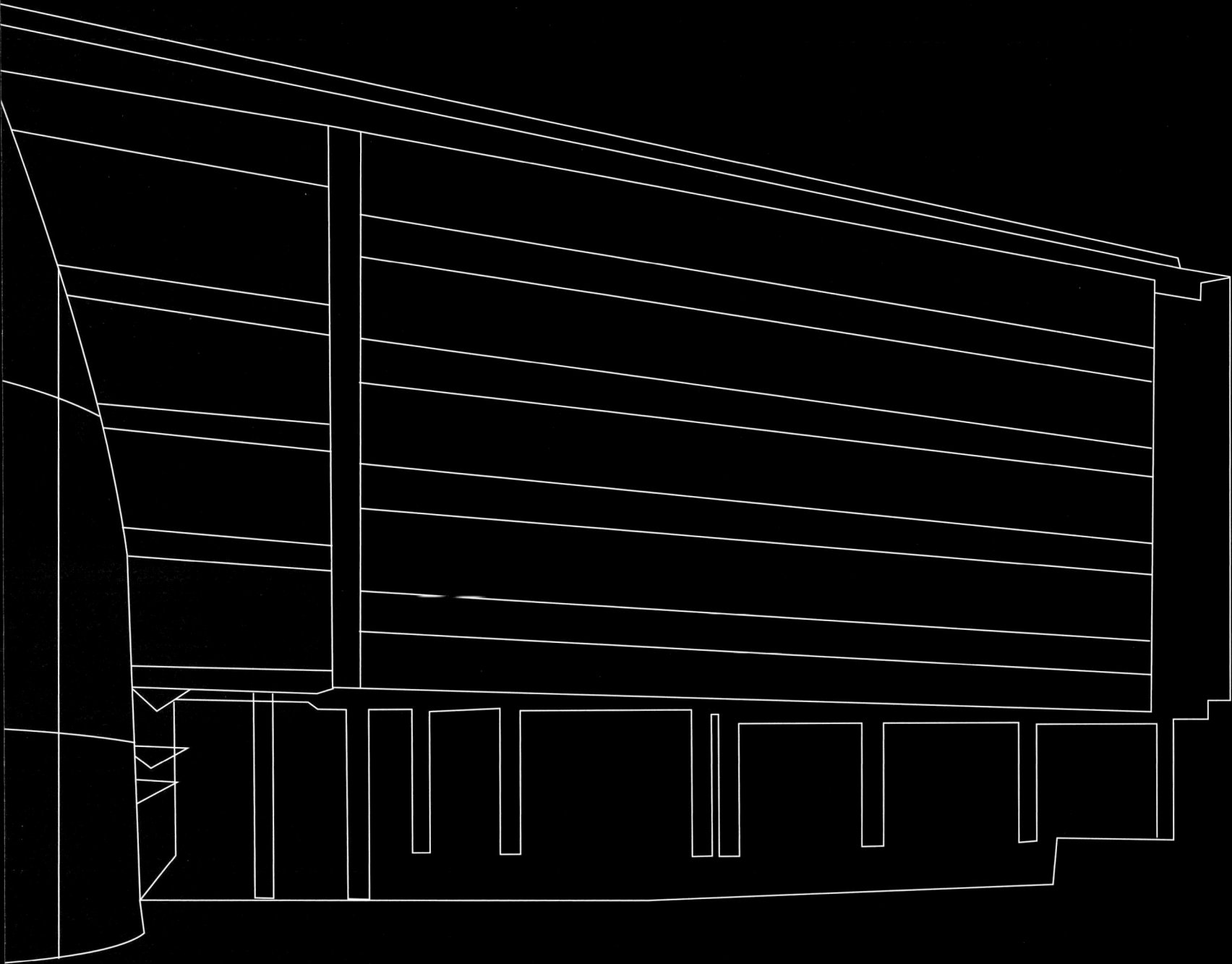

AP Møller School

Architect: Julian Weyer, Kessler&Kremer, Michael van Osten
Firm: C. F. Møller Architects
Location: Schlesvig, Germany
Area: 15000m²
Photographer: Julian Weyer, Kessler & Kremer, Michael van Osten, Poul Ib Henriksen

The AP Møller School is a Danish school in Germany. The aim of the project is to create a school which, in a straightforward manner, unites dreams and demands, community and individuality. The architectural expression of the school consists of two simple elements which, in their separate ways, combine sculptural form and function. The first of these is the overall brick body of the storeys, which possesses a dynamic expression with perforations and displacements in the facade. The second element is the extensive sloping copper roof surface which bisects the school body, revealing the school's structure and the contours of the two main inner spaces. The copper roof takes the form of an expansive, unifying gesture.

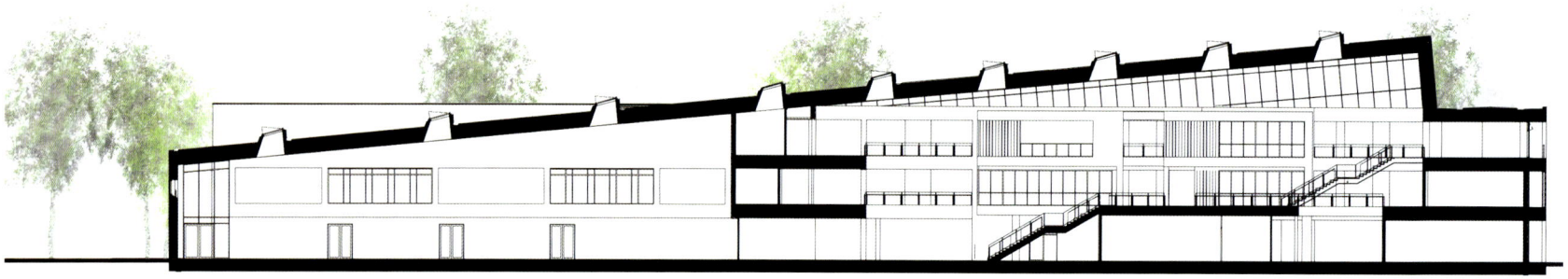

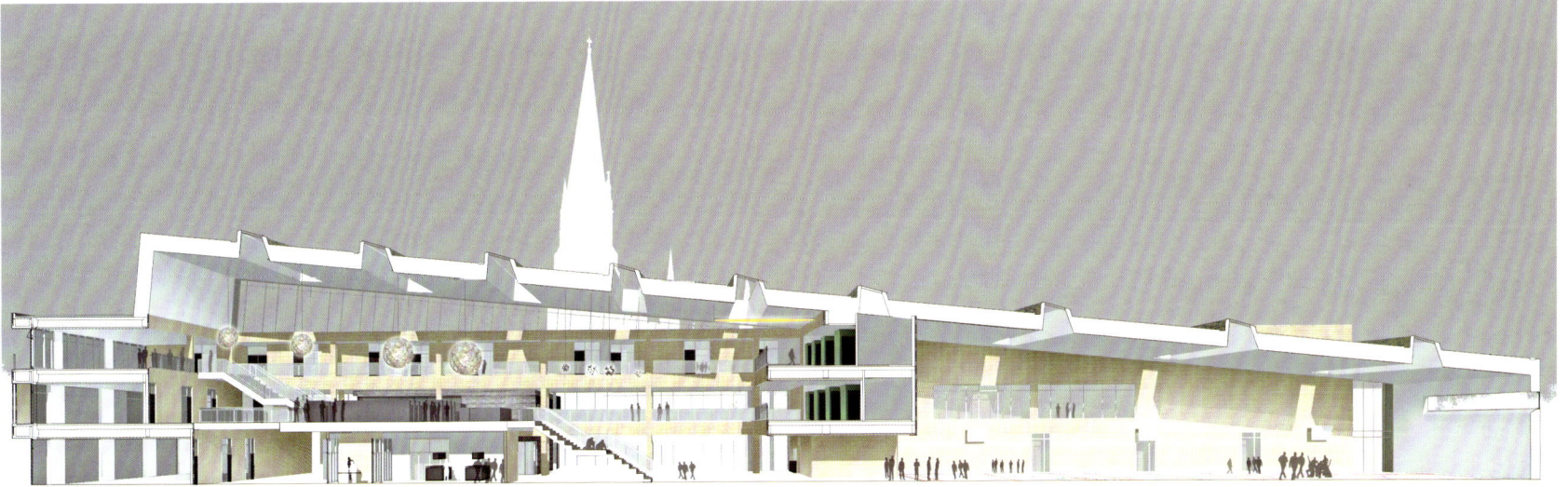

AP Møller学校是德国的一所丹麦学校。这个项目旨在直观地设计一个既梦幻又实用，既个性又能被大众所接受的校园。它的设计包括了两个简单的元素，分别在形式和功能上展现出来。第一个元素就是楼层的整体砖墙运用孔洞设计和错开排列的视觉感受，使整栋大楼呈现出一种统一的、动态的美感；第二个元素就是外观大面积的铜制屋顶的倾斜设计，使学校建筑一分为二，显露学校的主要结构和内部空间两个主体。广阔的铜制屋顶也呈现出校园整体统一的姿态。

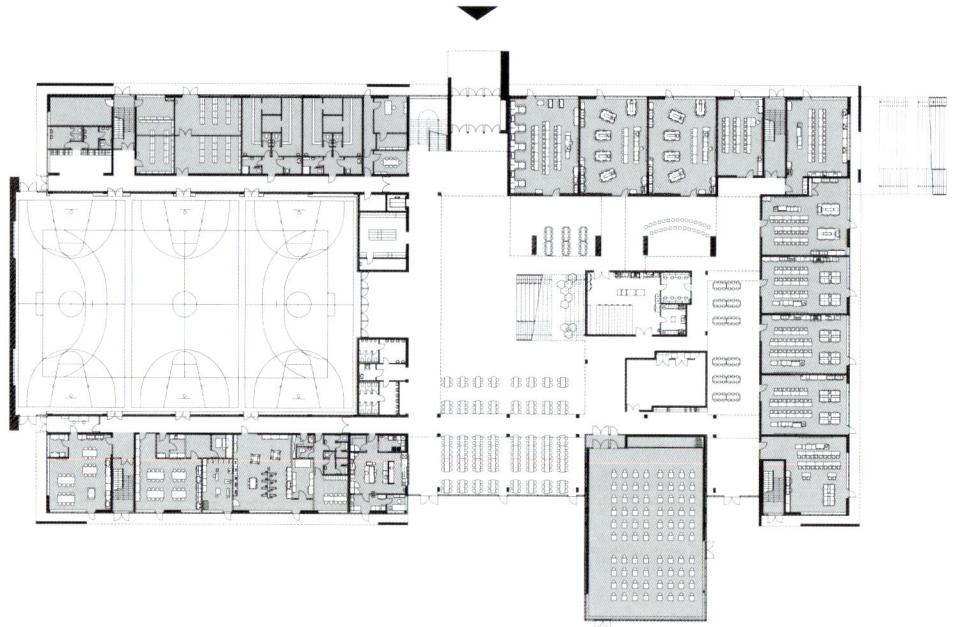

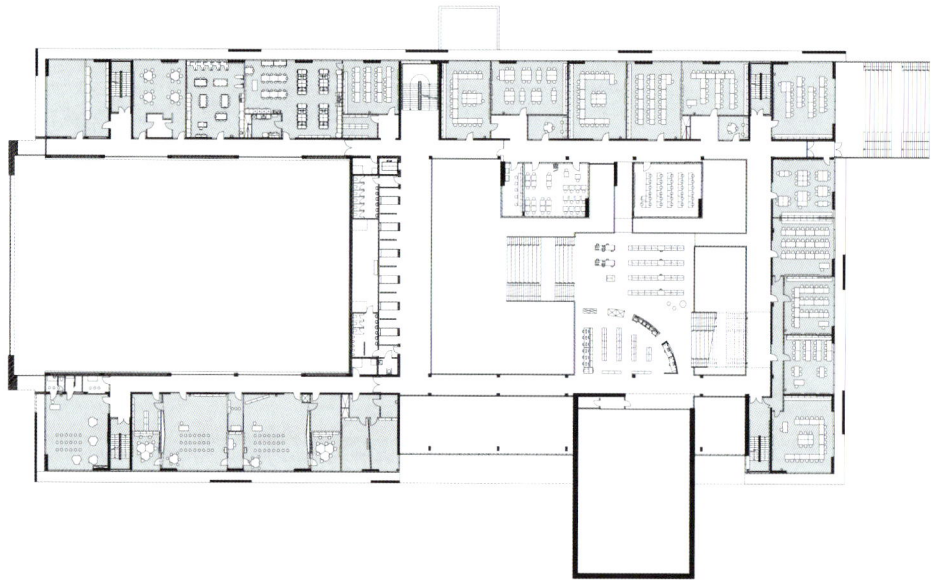

Faculty of Law, University of Sydney

Firm: Francis-Jones Morehen Thorp
Location: Sydney
Photographer: FJMT/Andrew Chung

The complex and extensive program for the project into podium and superstructures which allowed the creation of an extended public domain of lawns, terraces and plazas at the podium level, or Eastern Avenue level, the primary public artery of the campus. The material of the splintered forms that define this edge and opening of the University are layers of glass and timber plywood louvers, suspended on fine stainless steel rods. These forms possess a kinetic grain that changes with the position of the sun and preferences of those behind the timber screens. The pairing of large louver above small allows a variation in light and shade, blocking the harsh glare and heat of the low eastern and western sun while preserving and directing views when seated within.

The project has also developed from a focus upon the process of learning – combining the structured and unstructured, the formal and informal, the expected and unexpected, and the physical with the virtual – all within an uplifting environment characterized by fresh air, comfort and natural light that seeks to both enable and inspire students.

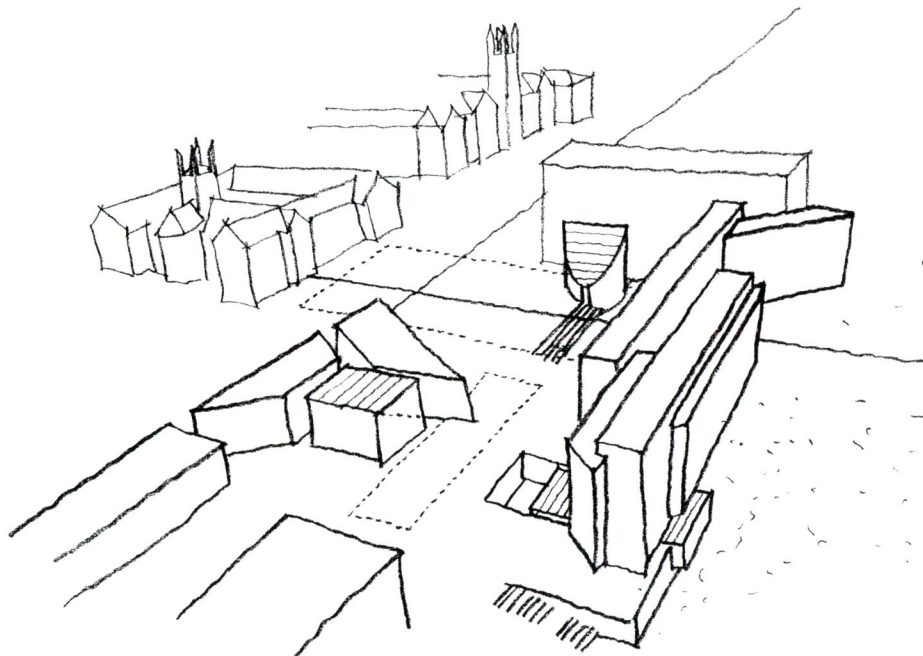

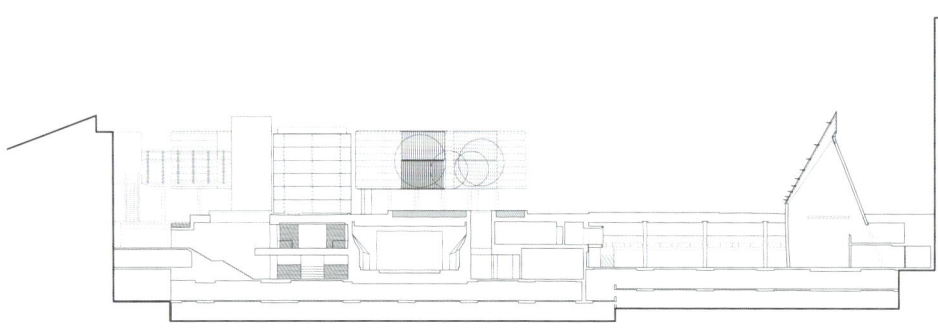

Faculty of Law
University of Sydney
Section through auditorium, library and light tower

0   5   10   20 m

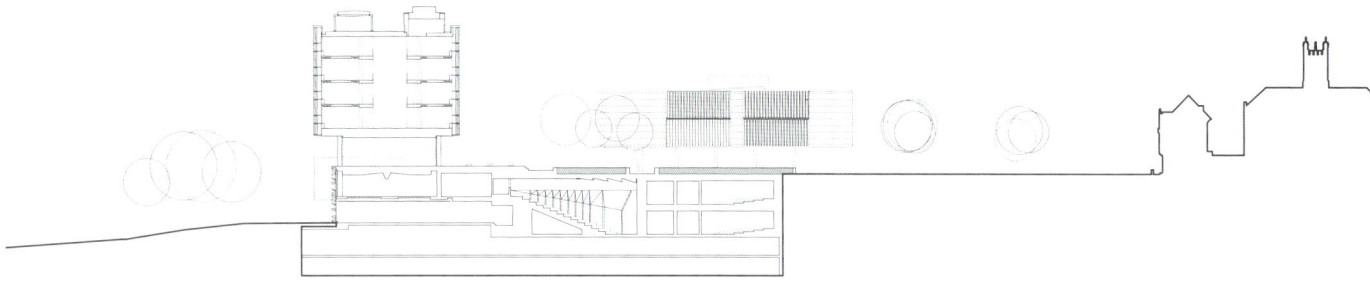

Faculty of Law
University of Sydney
Section through main lecture theatre

0   5   10   20 m

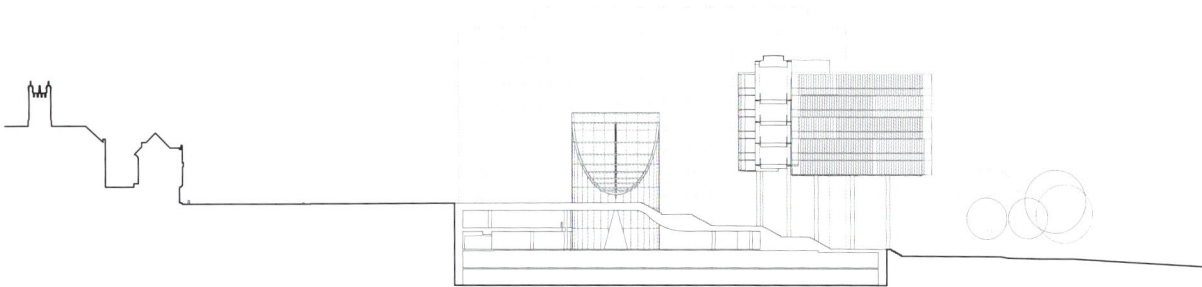

Faculty of Law
University of Sydney
Section through main stair

0   5   10   20 m

Faculty of Law
University of Sydney
Level 0 (Lobby Level)

Faculty of Law
University of Sydney
Level 1

Faculty of Law
University of Sydney
Level 2 (Garden Avenue level)

Faculty of Law
University of Sydney
Level 4 (Table 14 plaza)

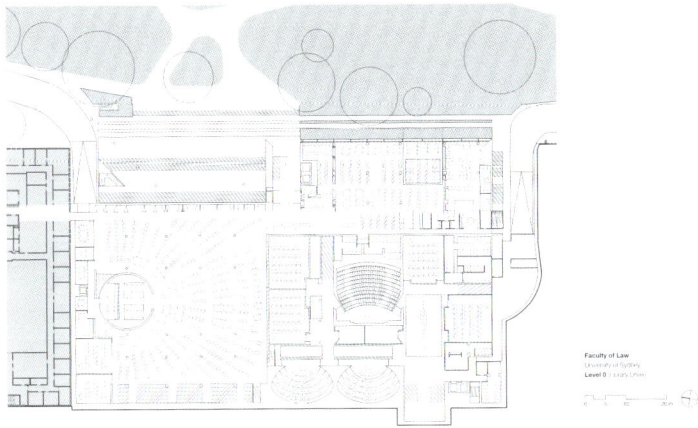

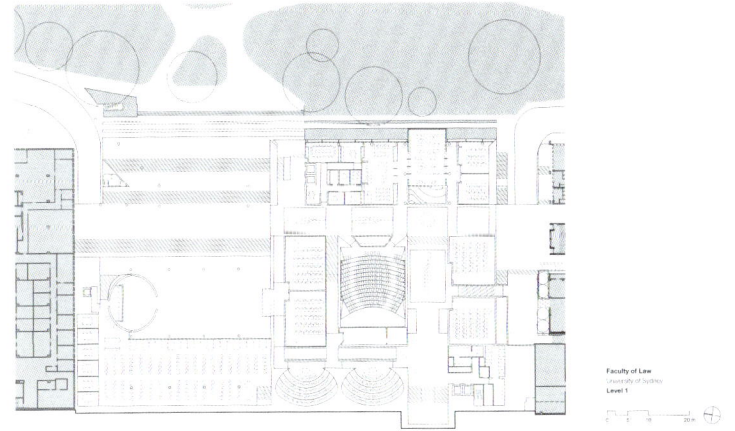

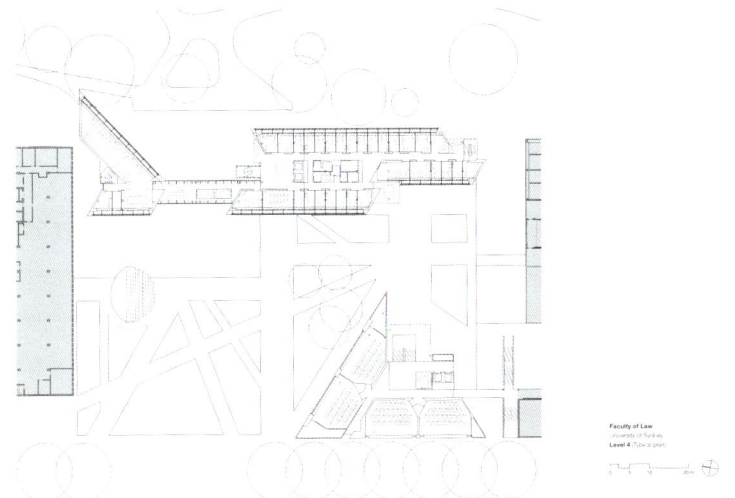

这个项目的复杂和广泛之处在于要在校园内建成一个允许创造能扩展公共领域草坪的拥有平台高度的露台和广场，或是具有东方的林荫大道水准的大学校园的主要校道。

用来界定大学的边界和开放的窗体的是多层玻璃和木质胶合板百叶窗，悬浮在优质的不锈钢棍上面。这些窗体拥有充满动感的纹理，它们能根据太阳的位置和后面的木质屏风的参数选择的变化而变化。以大百叶窗覆盖在小百叶上的配对方式，使光和影都能变化，既挡住了来自东方日出和太阳西下的刺目的眩光和热，又为坐在里面的人保留了开阔的视野。

这个项目也是以学习过程为焦点发展起来的：结合有组织的和无组织的，正式的和非正式的，预期的和未预期的，实体的和虚拟的，所有的这些都在一个拥有清新的空气、舒适的自然光线、令人振奋的环境中，力图启发学生，使他们能够更好地学习。

Vitus Bering Innovation Park

Architect: C. F. Møller Architects
Firm: C. F. Møller Architects
Location: Horsens, Denmark
Area: 8000m²
Photographer: Julian Weyer

Teaching and entrepreneur offices side by side – that's the philosophy behind the extension, which forms a distinctive addition to the existing structure.

The building's dynamic and innovative character is expressed via its spiral shape. On the facade, the movement is seen in the glazing strips that stretch towards the sky across the six storey of the building and allude a spiral, while internally it is expressed via the main stairway in green fiber cement, which runs in a spiral form between the storey in the unifying atrium. The stairway form also has the practical advantage of allowing a necessary fire escape route to be passed through the building.

The Vitus Bering Innovation Park is one of the first office complexes in Denmark to be classified as low-energy class 1, which means that its energy efficiency is twice that of the minimum required by the Danish building regulations. The low level of energy consumption is achieved through such factors as highly insulating windows and extra insulation on all of the building's external surfaces. Another feature is the building's intelligent air conditioning system, which adjusts itself according to the number of people present in each individual room.

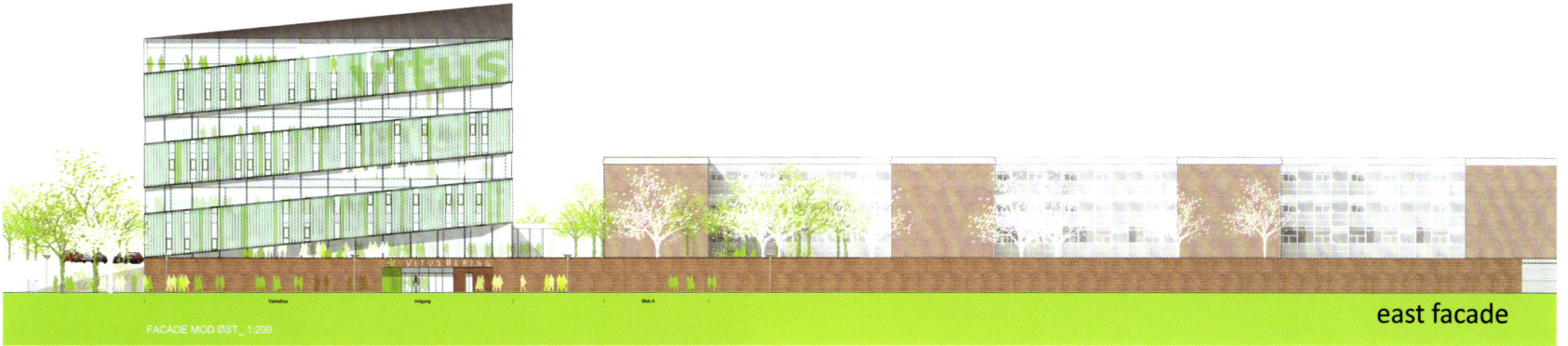

east facade

FACADE MOD ØST_ 1:200

south facade

FACADE MOD SYD_ 1:200

west facade

FACADE MOD VEST_ 1:200

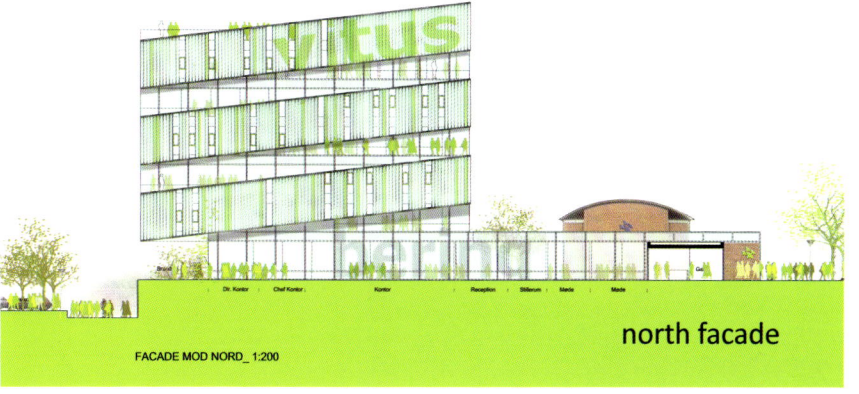

north facade

FACADE MOD NORD_ 1:200

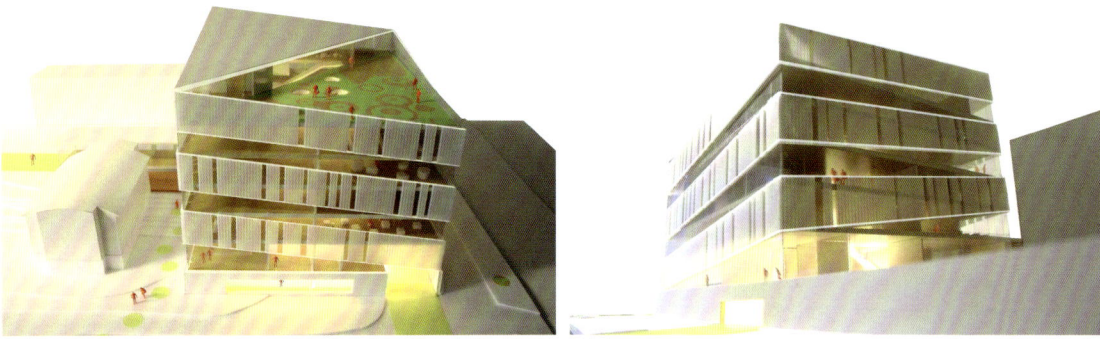

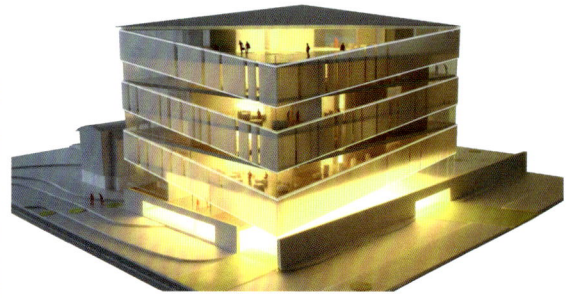

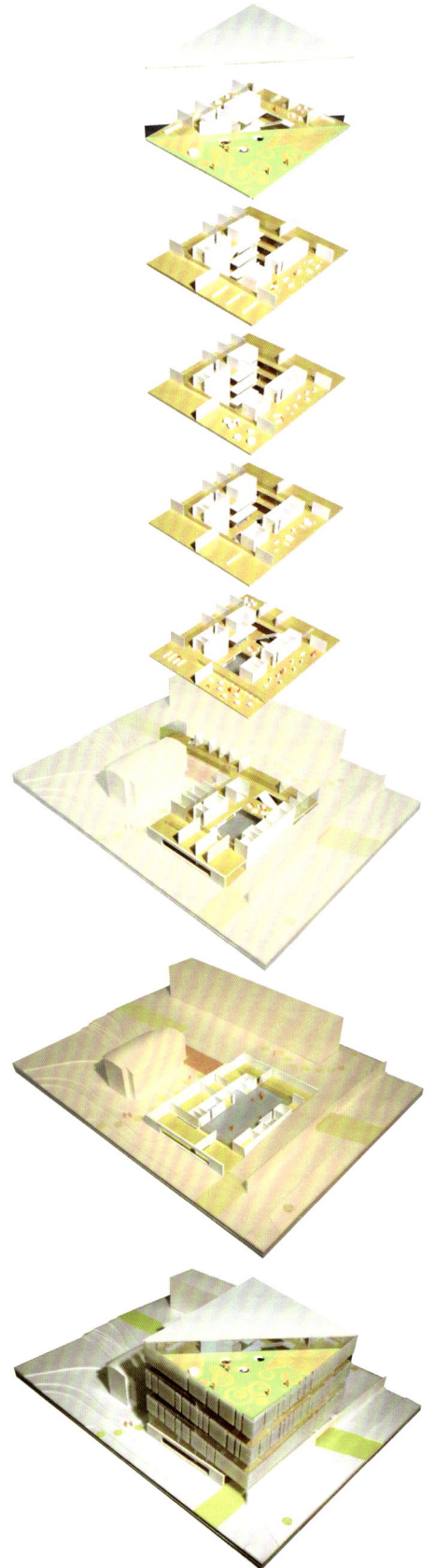

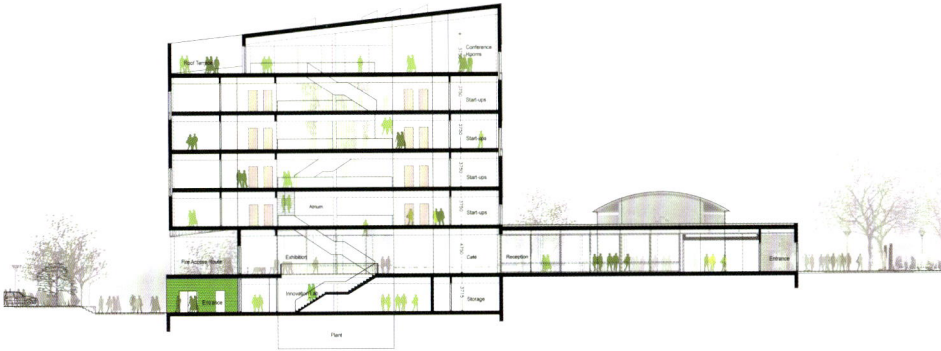

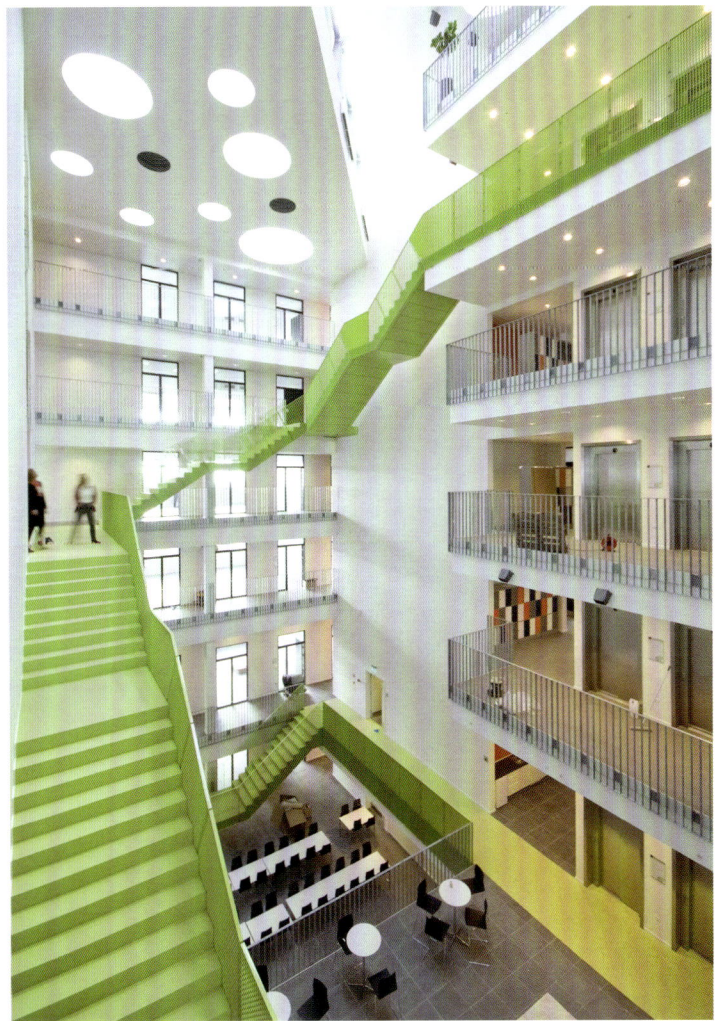

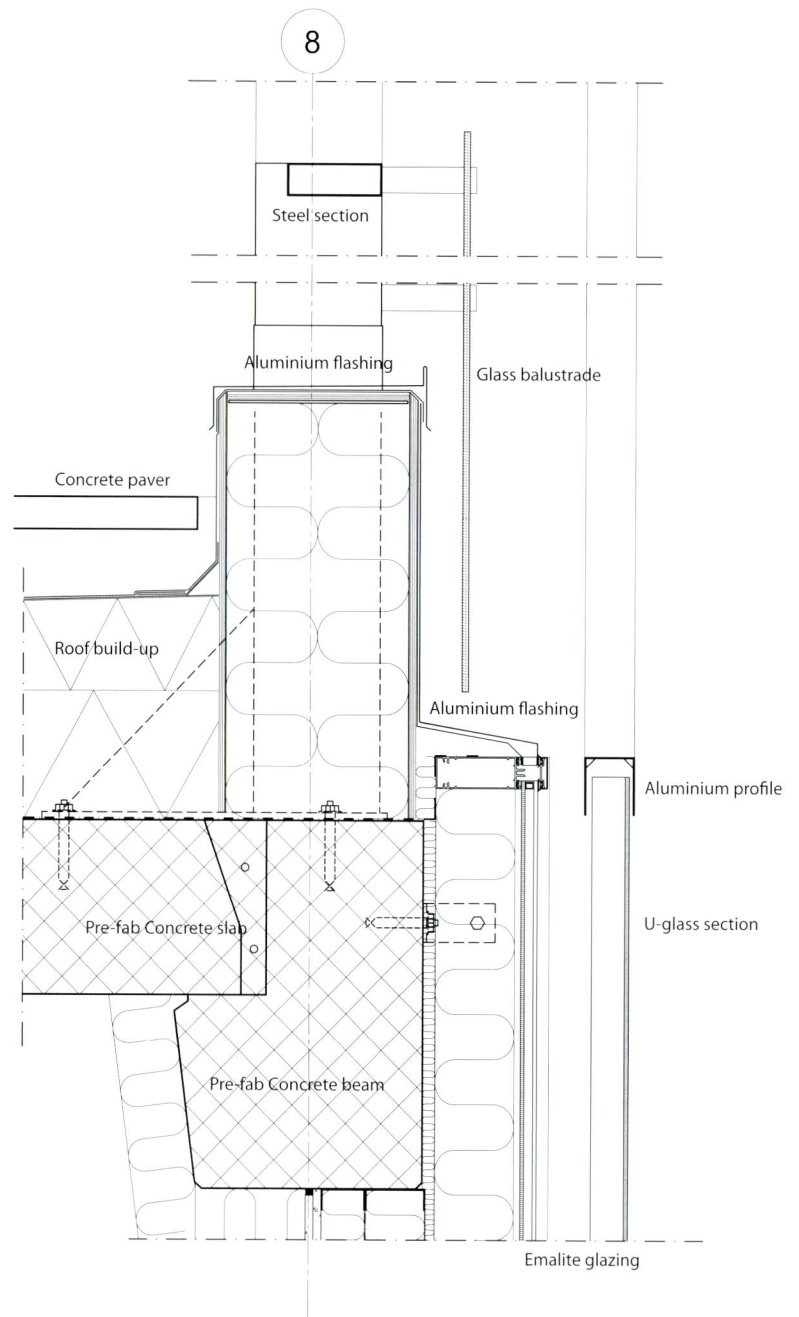

8

Steel section

Aluminium flashing

Glass balustrade

Concrete paver

Roof build-up

Aluminium flashing

Aluminium profile

Pre-fab Concrete slab

U-glass section

Pre-fab Concrete beam

Emalite glazing

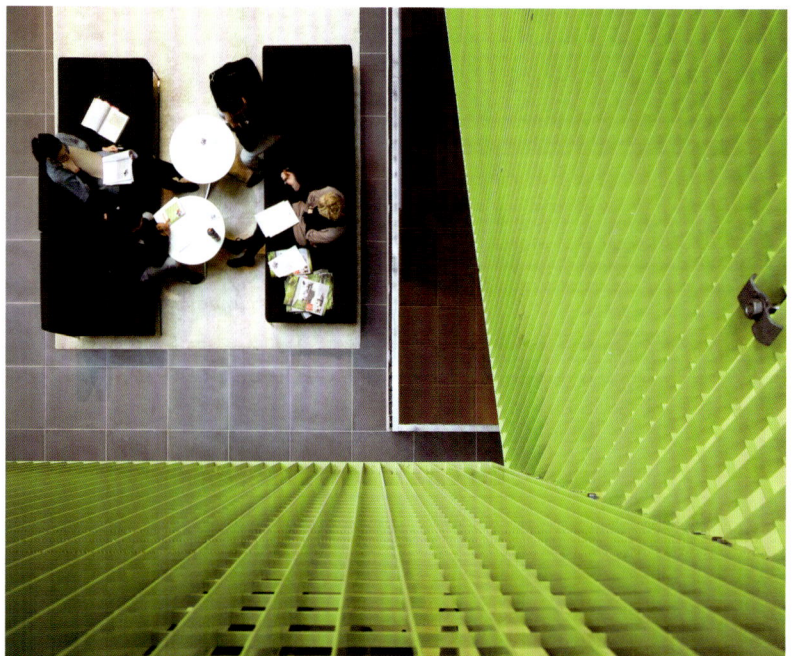

existing building

new plaza

new entrance

office

lounge

lounge

kitchenette

atrium

office

parking

existing auditorium

fire access route

fire access ramp

existing building

archives

archives

archives

archives

archives

entrance

lounge/ canteen/ workshop

office

office

fire access ramp

Møde 20-25 p. 40 m2

Mobil garderobe

Møde 10-15 p. 21 m2

Møde 10-15 p. 21 m2

Receptions område

Møde 20-25 p. 40 m2

Lounge/"Fatboy" område

Toilet 3 m²

Reng. 3 m²

Ophold

Terrasse

Møderum 6-8 p. 14 m²

Kryddsfelt

Møderum 6-8 p. 14 m²

Tekøkken 7 m²

Ovenlys

Møde 20-25 p. 37 m²

Skald

Ovenlys

Terrasse

Ophold

TAGLANDSKAB

lounge

Lounge"Fatboy" område

teaching

Undervisning 2x21 p. 150 m²

4-6-mandskontor 47 m²

office

kitchenette

atrium

exhibition

kitchenette

lounge

teaching

office

威图斯·白令创新园是一座教学工作场所与创业启动办公设施相结合的办公大厦。新创新大楼在原有的结构设计上直接延续了原有综合楼的建筑风格，但是新楼的与众不同仍然是显而易见的。

建筑充满活力和创新的特色通过螺旋造型表现出来。立面上的玻璃窗带环绕着六层高的建筑攀升，带来一种螺旋接续的印象，同时楼内也有一条绿色纤维水泥主楼梯，顺着中庭以螺旋状向上衔接起每个楼层。建筑的倾斜外形也具有实用价值，它允许必要的防火梯穿过建筑。

威图斯·白令创新园是丹麦低能耗水平为一级的最早的综合办公楼之一，这意味着其能效是丹麦建筑条例要求的最小值的两倍。低水平的能耗通过如下因素获得：保温效果极佳的窗户和在所有建筑外表皮上添加额外保温层。建筑的另一大特色是智能空调系统，它能根据每个房间的人数来自动调节功率。

Orestad College

Architect: Kim Herforth
Firm: 3XN
Location: Orestad, Copenhagen, Denmark
Area: 12000m²
Photographer: Adam Mork

Orestad Gymnasium (Orestad College) is the first college in Denmark based on the new visions of content, subject matter, organization and learning systems. Communication, interaction and synergy have been key issues.

The project displays a visionary interpretation of openness and flexibility regarding team sizes, varying from the individual over groups to classes and assemblies, and reflects international tendencies aiming at achieving a more dynamic and life-like study environment and introducing IT as a main tool.

The college is interconnected vertically and horizontally. Four boomerang shaped floor plans are rotated to create the powerful super structure which forms the overall frame of the building – simple and highly flexible. Four study zones occupy one floor plan each. Avoiding level changes makes the organizational flexibility as high as possible, and enables the different teaching and learning spaces to overlap and interact with no distinct borders. The rotation opens a part of each floor to the vertical tall central atrium and forms a zone that provides community and expresses the college's ambition for interdisciplinary education.

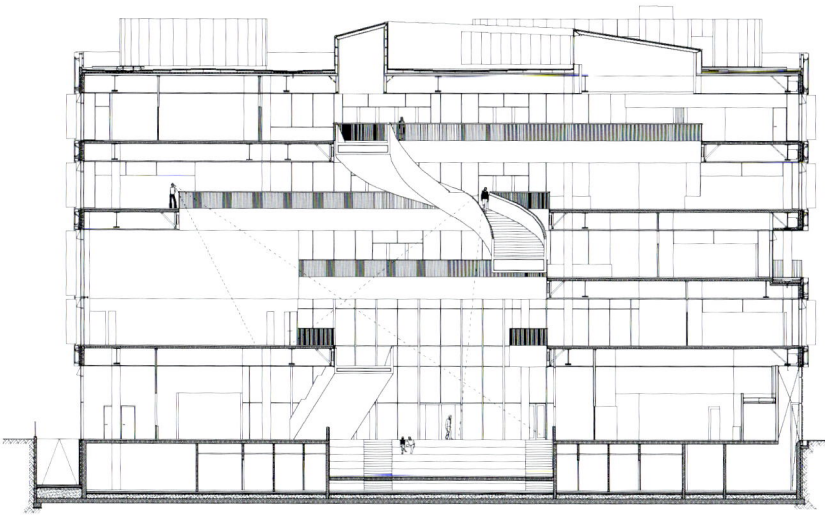
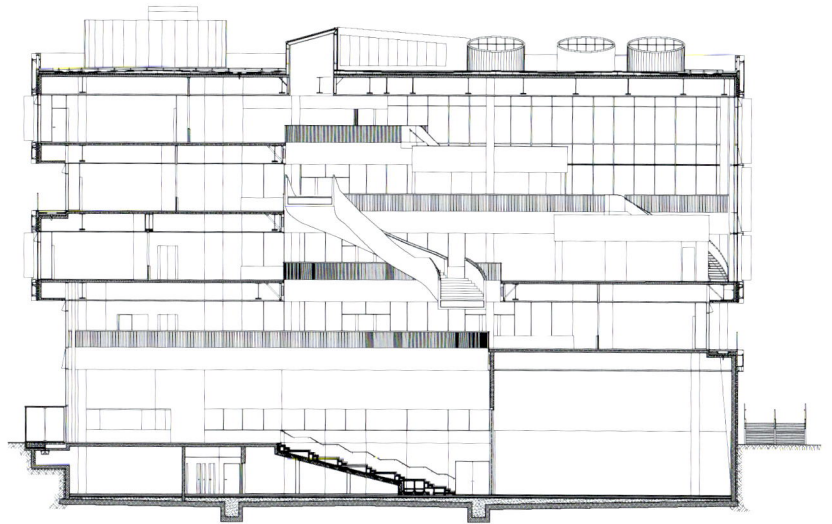
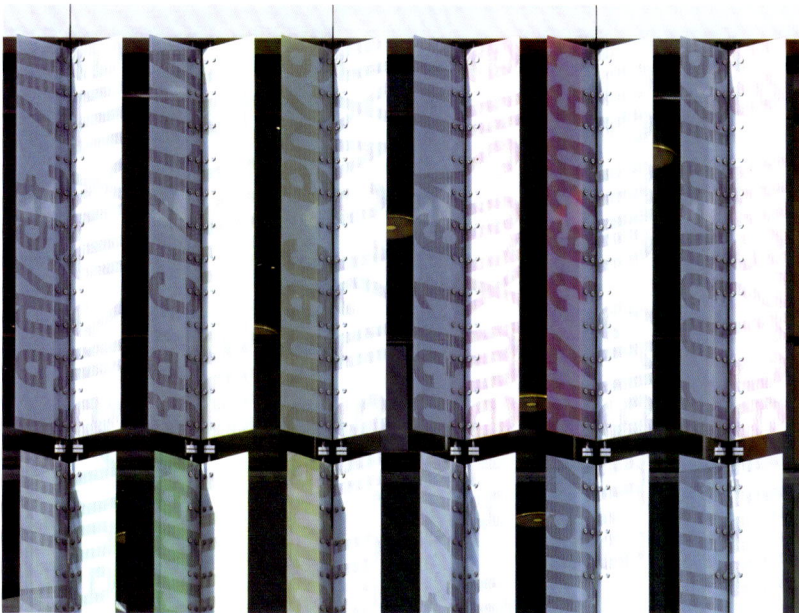

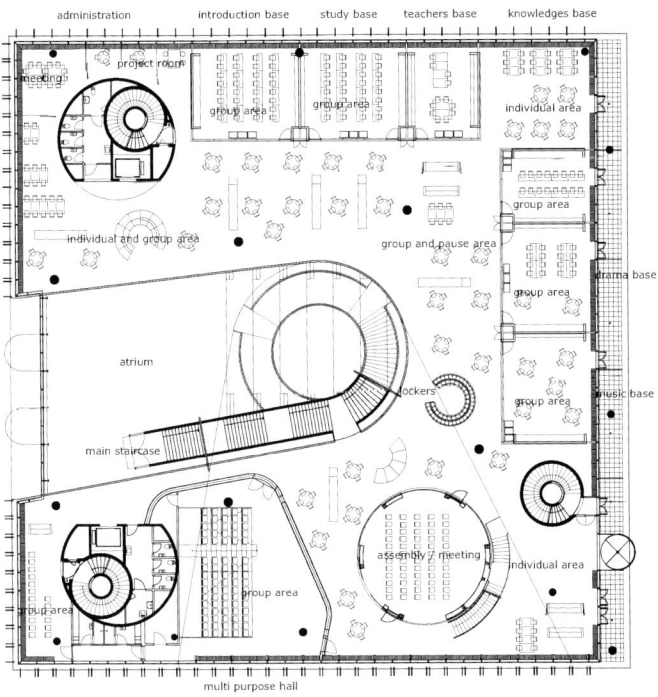

administration   introduction base   study base   teachers base   knowledges base

meeting   project room

group area   group area   individual area

individual and group area   group area

group and pause area   drama base

atrium   lockers   group area   music base

main staircase

group area   group area   assembly / meeting   individual area

multi purpose hall

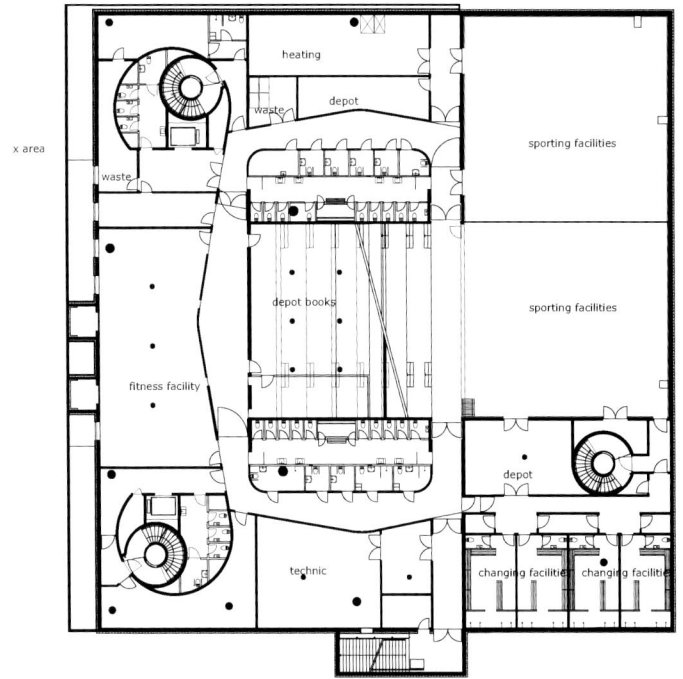

heating

x area   waste   depot   sporting facilities

depot books   sporting facilities

fitness facility

depot

technic   changing facilities   changing facilities

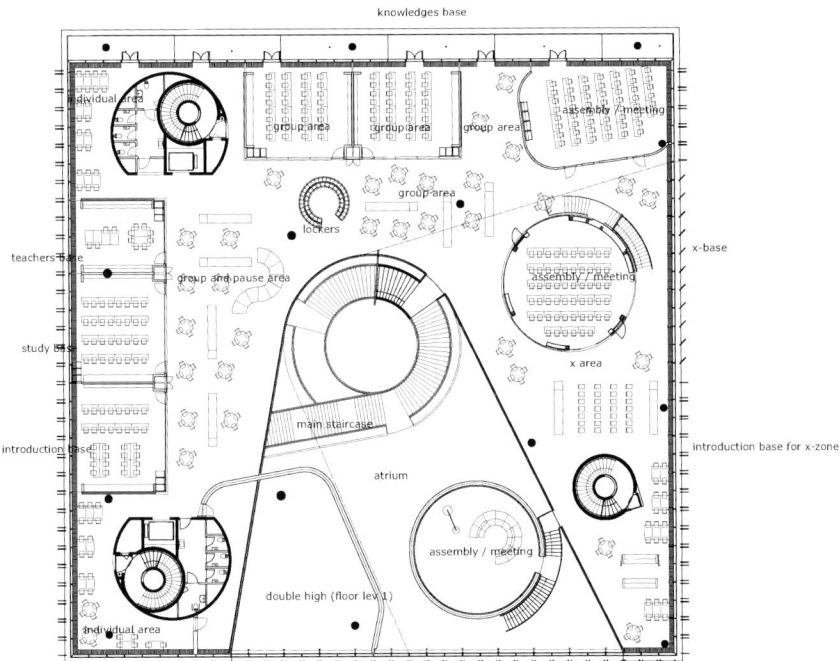

knowledges base

individual area   group area   group area   group area   assembly / meeting

group area

teachers base   lockers

group and pause area   assembly / meeting   x-base

study base   x area

introduction base   main staircase   introduction base for x-zone

atrium   assembly / meeting

double high (floor level 1)

individual area

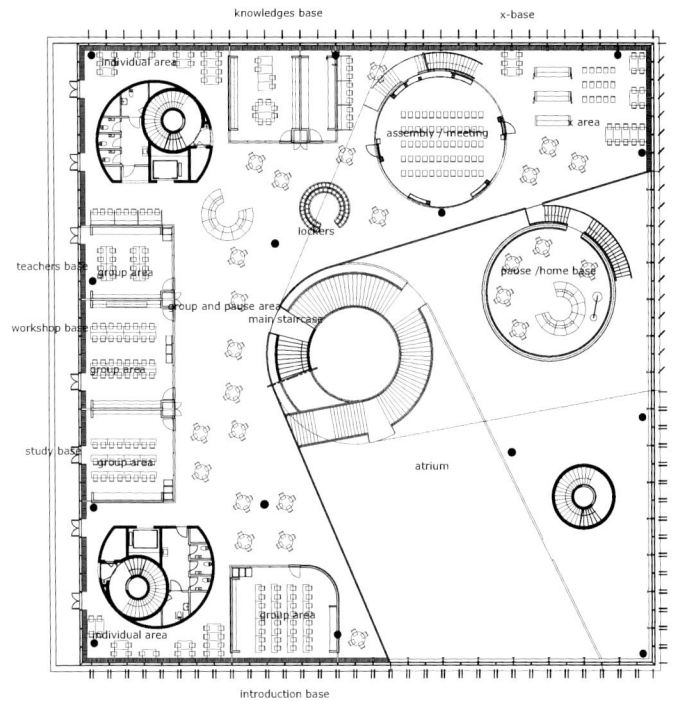

knowledges base   x-base

individual area   assembly / meeting   x area

teachers base   lockers   pause /home base

workshop base   group and pause area   main staircase

group area

study base   group area   atrium

individual area   group area

introduction base

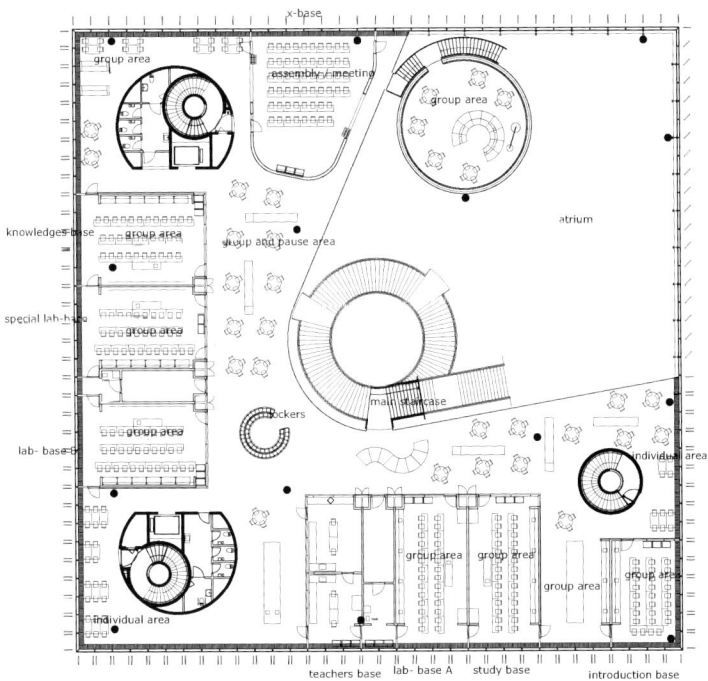

x-base

group area   assembly / meeting   group area

atrium

knowledges base   group area   group and pause area

special lab-base   group area

lab- base B   lockers   main staircase   individual area

group area

individual area   group area   group area

teachers base   lab- base A   study base   introduction base

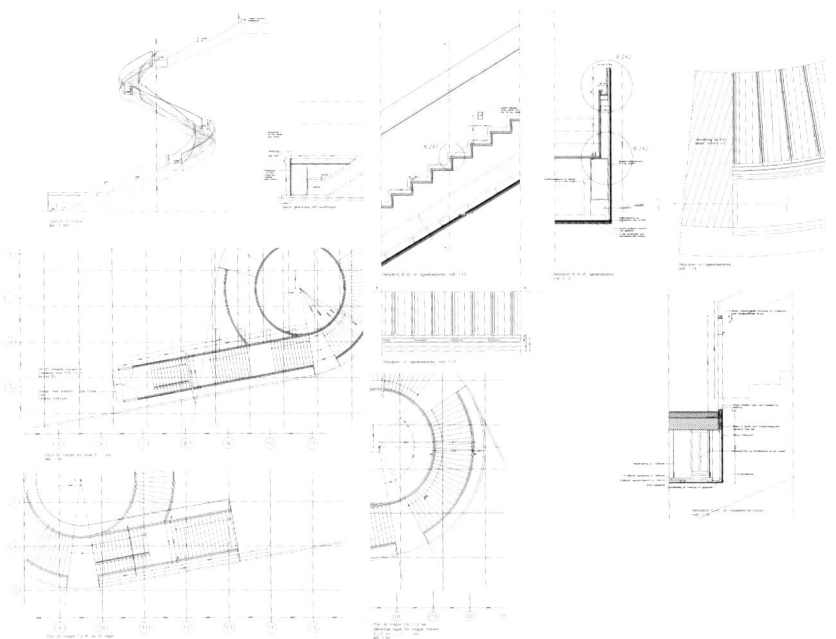

Orestad Gymnasium 是丹麦第一个以创新教学内容、创新课程设置、创新组织形式和创新学习模式为创办基础的学校。沟通、互动和协作是教学的重要课题。

这个项目向人们展示了一个开放灵活的团队，无论是在个人还是在班集体或其他集体上，体现了实现动感、轻松的学习环境和运用IT作为主要教学工具的国际潮流。

大学校园的设计纵横交错。用四个螺旋楼梯来构建大楼的主要构架，既简单又具动感。四个学习区分布在四个角。有效避免了水平变化，使得组织构架看起来更加灵动，也使得不同的教学区和学习区重叠交错，没有很明显的界限。这种每一层楼的盘旋交错形成了一个高的垂直中庭，也提供了一个交流社区，表达了这所学校追求一种学科间互相渗透的学习模式。

Ciss High School

Firm: Perkins Eastman
Location: Shanghai, China
Area: 12720m²
Photographer: Copyright ShuHe
Drawings: Courtesy Perkins Eastman

The building's composition attempts to break all visual cues typically associated with traditional school design. Echoing—but not replicating—the visual vocabulary of previous phases, building materials are reinterpreted with an increasingly sophisticated and distinguished treatment.

The largest portion of the green roof is landscaped over light monitors located above the gymnasium/events space. This habitable garden, which uses soil made from recycled crushed bricks, is located with direct access from the adjacent science suite for cross-curricular experiments and the incorporation of ongoing field studies conducted by students on weather and air quality—making it an important part of curriculum. Other environmentally responsible features and materials include day lighting, natural ventilation, and materials include bamboo and natural rubber flooring.

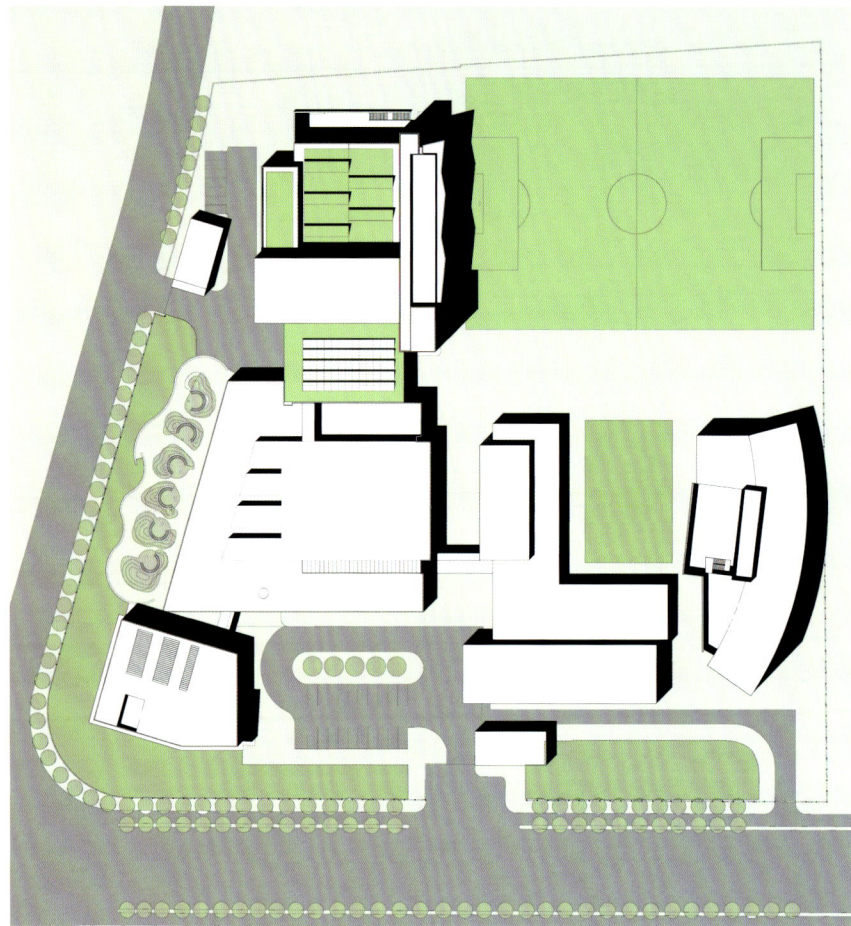

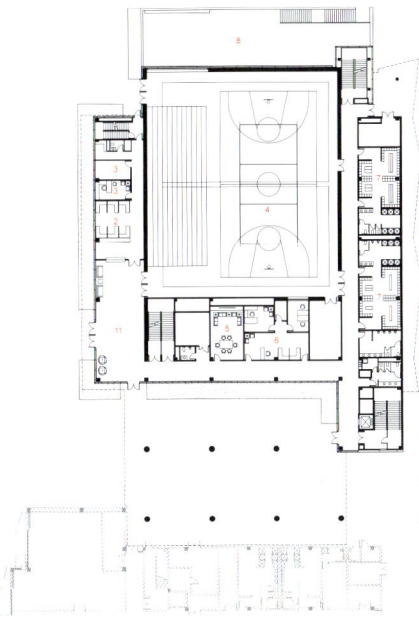

CISS HIGH SCHOOL, floor one

1 Lobby
2 Open office area
3 Office
4 GYM
5 Conference room
6 PE office
7 Locker room
8 Sunken garden
9 Existing phase –

0 5 10   25 METERS

CISS HIGH SCHOOL, floor two

1 Classroom
2 Teacher's work room
3 Small group instruction
4 Guidance
5 Open office
6 Office
7 Conference
8 Library
9 Work room
10 Information commons
11 Internet cafe
12 Open to GYM below
13 Existing phase –

0 5 10   25 METERS

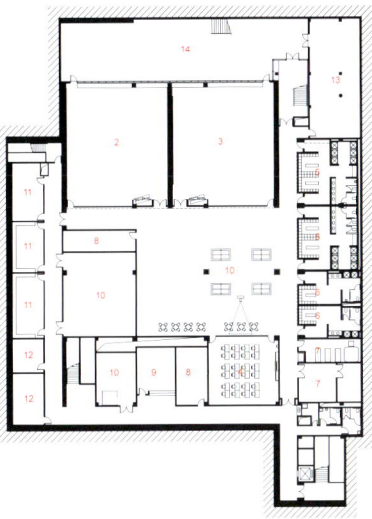

CISS HIGH SCHOOL, basement floor

1 PE commons
2 Fitness
3 Dance
4 Team classroom
5 Team locker room
6 Staff locker room
7 Training
8 Storage
9 Copy center
10 Mechanical
11 Av room
12 Electrical
13 Pump room
14 Garden

0 5 10   25 METERS

CISS HIGH SCHOOL, floor five

1 Classroom
2 Teacher's room
3 Science
4 Wet lab
5 Preparation

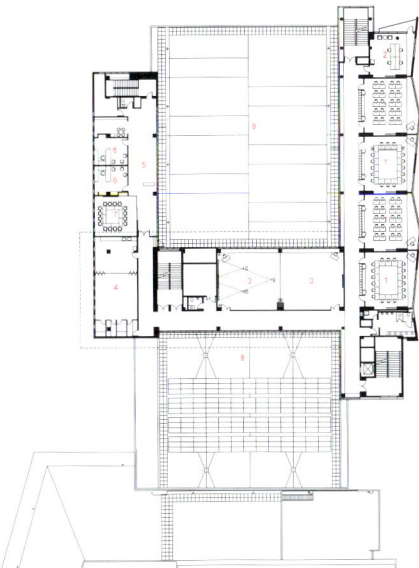

0 5 10   25 METERS

CISS HIGH SCHOOL, floor three

1 Classroom
2 Teacher's work room
3 Large group instruction
4 World language lab
5 Open office
6 Office
7 Board room
8 Green roof

0 5 10   25 METERS

CISS HIGH SCHOOL, floor four

1 Classroom
2 Teacher's work room
3 Science
4 Wet lab
5 Preparation
6 Science Garden

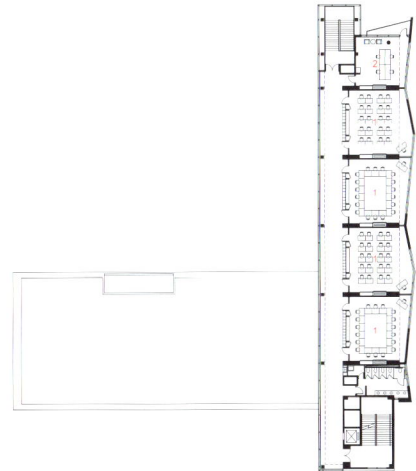

0 5 10   25 METERS

CISS HIGH SCHOOL, floor six

1 Classroom
2 Teacher's room

0 5 10   25 METERS

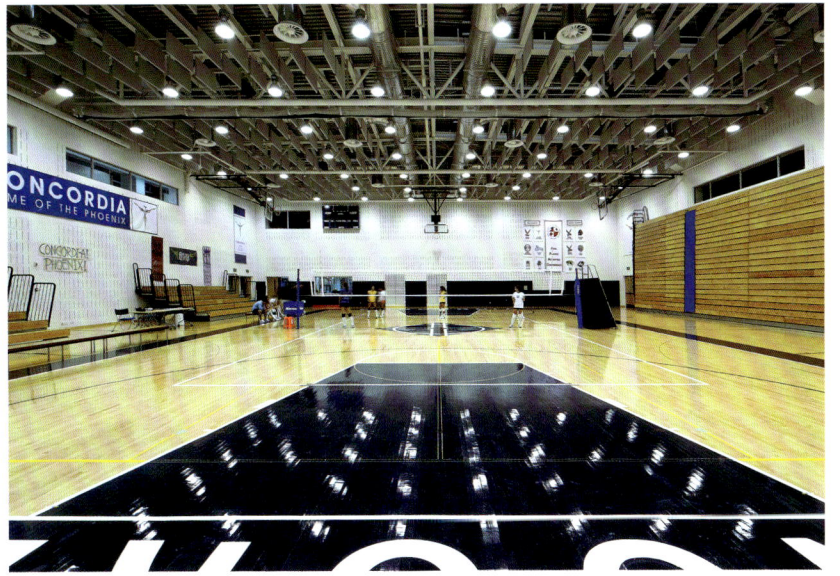

这个作品的设计在于打破传统的学校设计的视觉效果。与往往是复制模仿的传统学校设计的视觉感受不一样，这一次的设计旨在引起共鸣，用日益成熟和独特的眼光重新感受建筑材料带来的感觉。

最大面积的绿化屋顶底下是灯光控制器，这个控制器位于体育馆上方。在实验室旁边有一个花园，花园里的环境十分宜人。泥土是用压碎的石子再制而成，循环利用，这个花园也可以供学生进行天气和空气质量的课外实验和实地研究等重要课程。其他重要的功能特性包括采光、自然风，使用材料包括竹子和天然橡胶等。

The Langley Academy

Architect: Foster and Partners, Nigel Dancey, Iwan Jones, Dominik Hauser, Rachel MacIntyre, Adrian Nicholas, Declan Sharkey, Stefan Unnewehr, Pietje Witt
Location: Slough
Area: 10064m²

The main entrance to the Academy is a large, open, flexible atrium, suitable for informal gatherings and assemblies. It is a generous exhibition space for loaned museum exhibits and schoolwork, and has direct views into the laboratories and domestic science lab. Pupils will be encouraged to learn about curating exhibitions by having direct experience of putting displays together. The plant room, ducts and pipes are revealed and energy use is available for scrutiny, and students are drawn into a proactive drive towards achieving a sustainable building. The atrium has views into the restaurant and through to the playing fields, so that the sporting achievements of the school are integrated and become key to the collective sense of pride and community.

Langley is a form-based academy, for pupils between the ages of 11 to 18, and classrooms are configured around two "fingers", which extend from the entrance atrium. The classrooms are highly flexible teaching spaces, containing ergonomic furniture, perimeter teaching facilities and projection screens on all four walls. The idea is that these spaces can function like conventional classrooms as well as open, adaptable rooms. This enables out of hours use of the building for the community, as well as the option for teachers to use less formal teaching methods.

Langley Academy - Site Plan

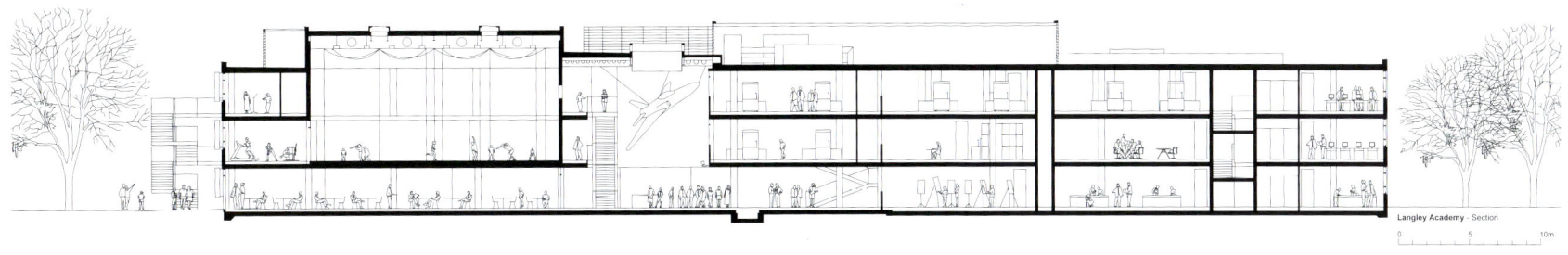

Langley Academy · Section

0   5   10m

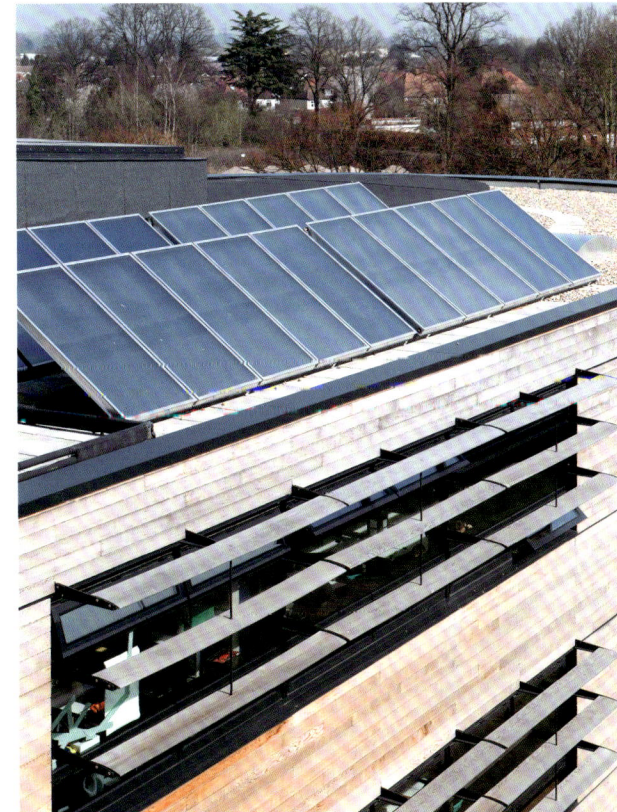

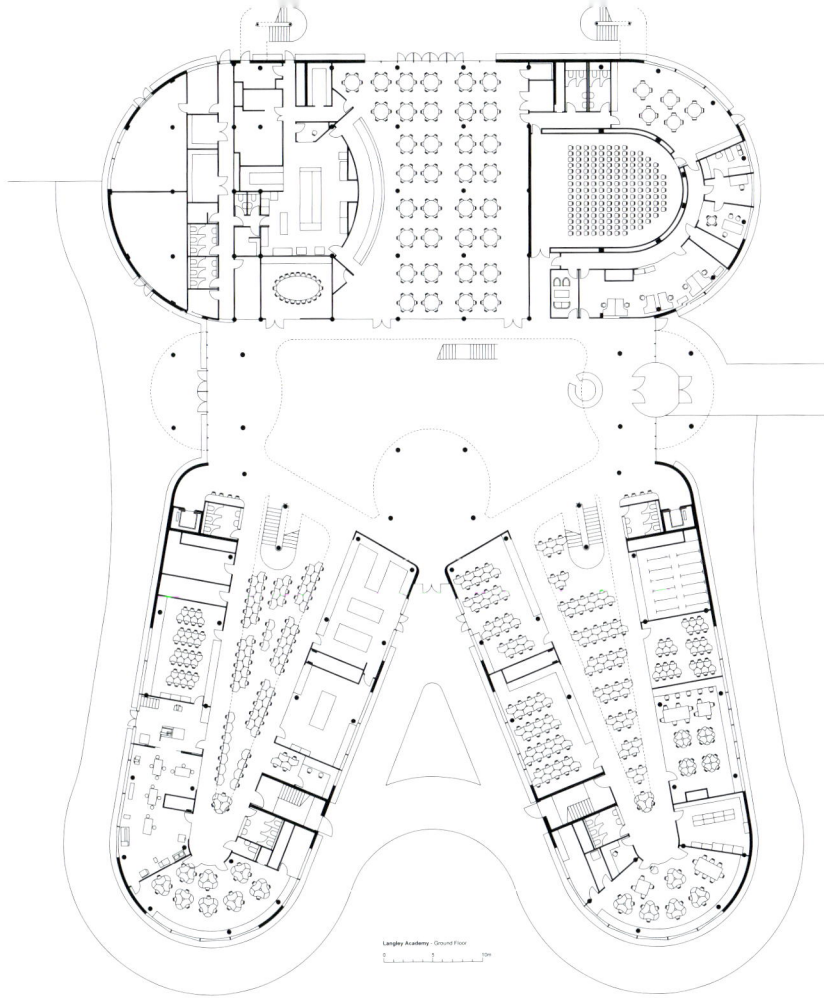

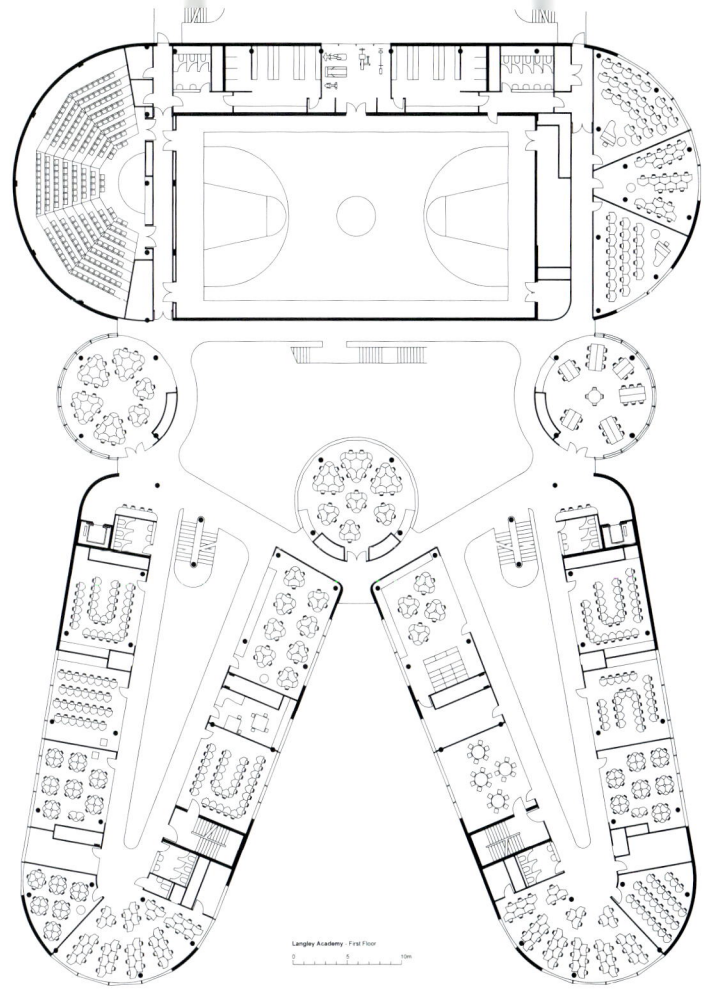

Langley Academy - Ground Floor

Langley Academy - First Floor

　　这个学院的主要入口是一个宽大、灵活而开放的中庭，适合于正式和非正式的集会。作为可用于外借举行博物馆展览和学校活动的场所，这算是一个宽敞的展览空间，还能看见普通实验室和家政学实验室。学生们被鼓励通过直接展览的经验来学习关于策划展览的知识。在机房，通风管道和能源消耗都能详细检查，学生可以前瞻性地实现可持续发展的建筑。这个中庭还能眺望食堂和运动场，所以设计把展览和运动项目结合在一起，成为建立学校的荣耀和团结意识的关键场所。

　　兰勒是一个以形式为基础的学院。学生的年龄介于11到18岁之间。从中庭延伸出来的教室形状仿如两根手指。这些教室是高度灵活的教学空间，包括符合人类工程学的家具、教学设备和投影屏幕。这些空间既可用来当教室，也可用作开放的场所，在课余时间用于社团活动，也能便于教师进行户外的教学活动。

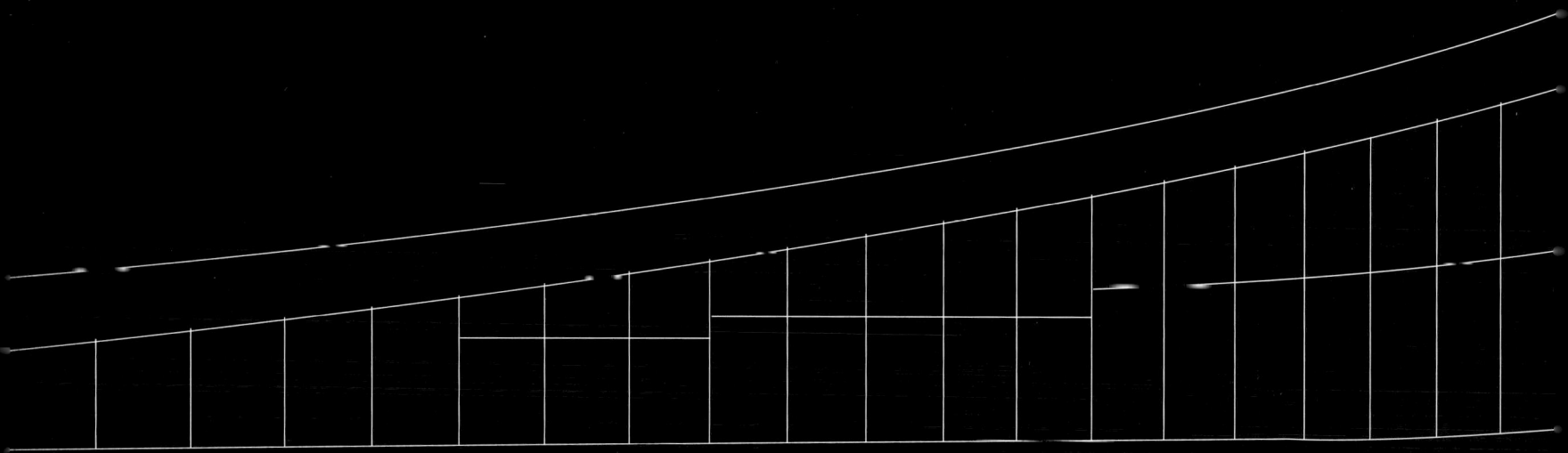

Entertainment & Leisure

娱乐与休闲

Bubbletecture H

Architect: Endo Shuhei
Firm: Endo Shuhei Architect Institute
Location: Sayo-cho, Hyogo Pref.
Area: 5000m²
Photographer: Yoshiharu Matsumura

- -

The client requested that all people who will visit this place including the inhabitants of Hyogo prefecture, improve the interest for global environmental concerns and are able to experience various approaches as the place of environmental study. The designer thought about creating the new environment architectural space that could share the point of contact with nature and environment providing a keyword called "the circulation" in a relation with nature for the request.

The site is on the steep slope of the north side in the forest. After integrating all functions that the client requested at the design term into three, required area and volume were set according to the function and structure. The two functions were arranged in parallel on flat land (old town road was there) that had remained at a high level in the site, and another was arranged having floated from the slope at the same level of other two. This is because it is assumed very important to make use of the limited flat land, keep natural landform as large as possible, and to minimize the influence of construction against peripheral natural environment. The form of the building shows the rational shape that connects these three functions for plane and section.

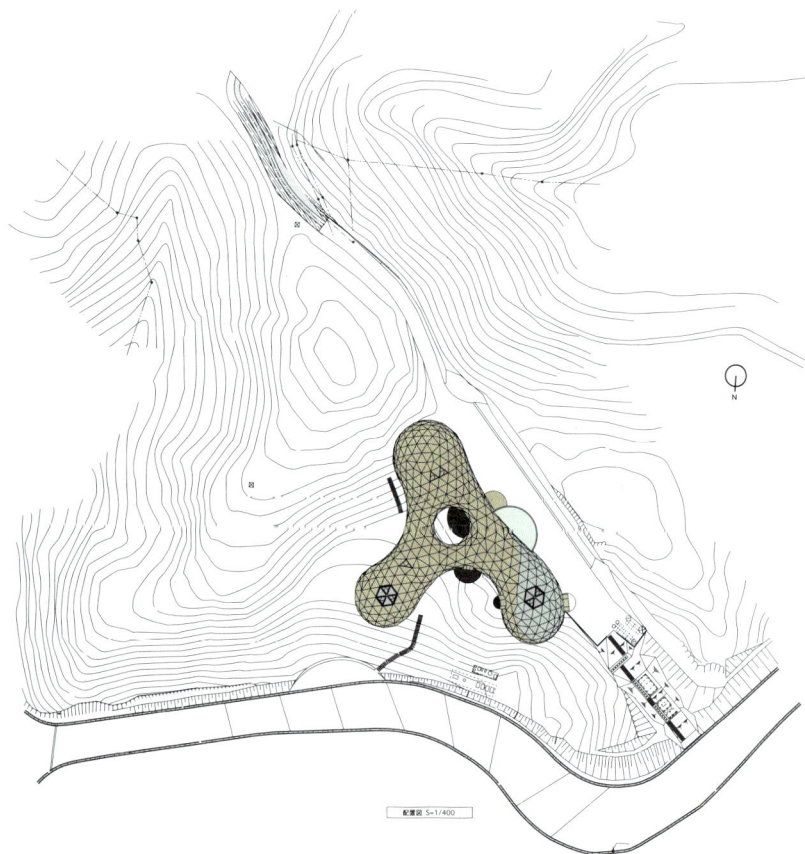

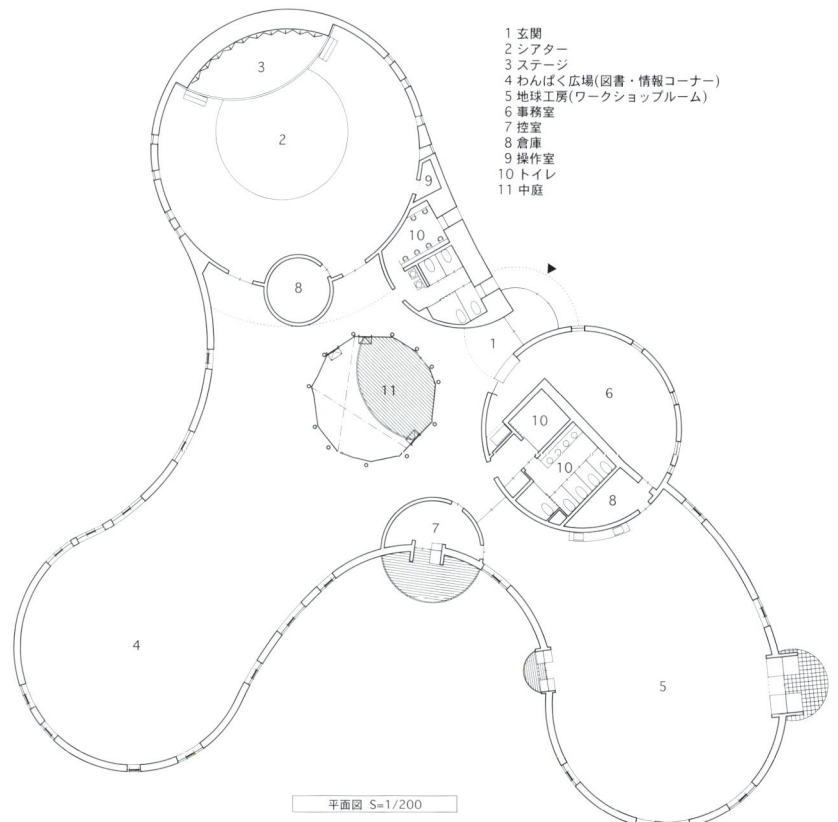

1 玄関
2 シアター
3 ステージ
4 わんぱく広場(図書・情報コーナー)
5 地球工房(ワークショップルーム)
6 事務室
7 控室
8 倉庫
9 操作室
10 トイレ
11 中庭

平面図 S=1/200

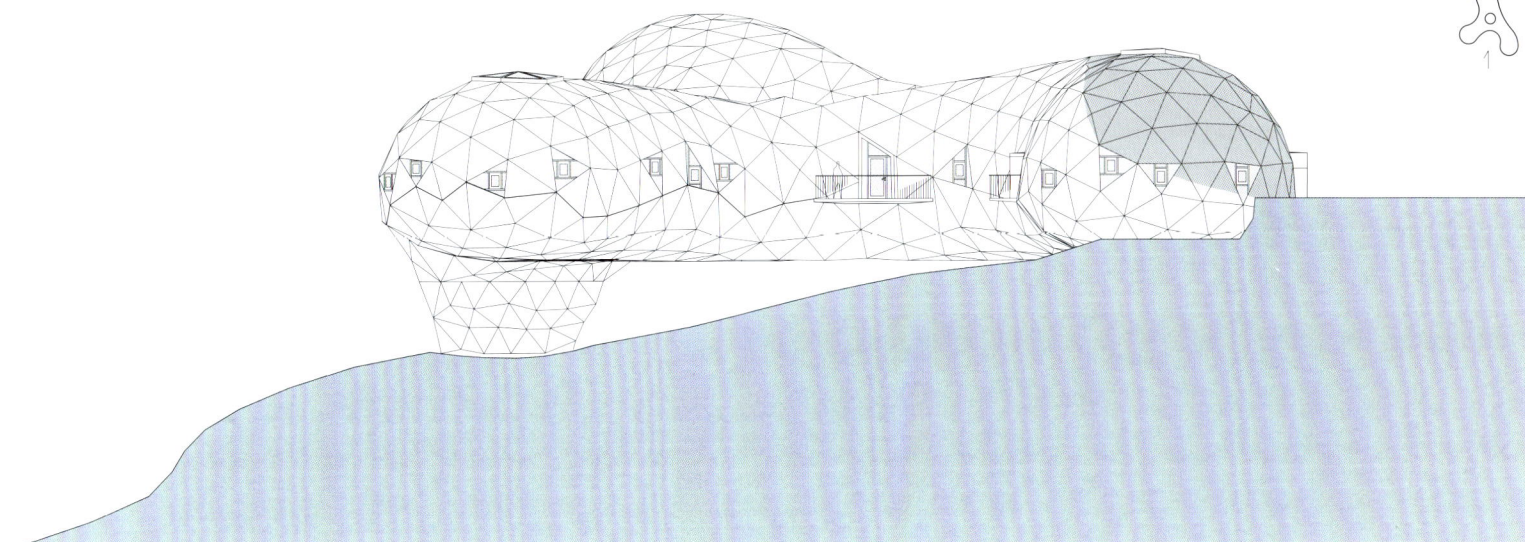

北立面図 S=1/200

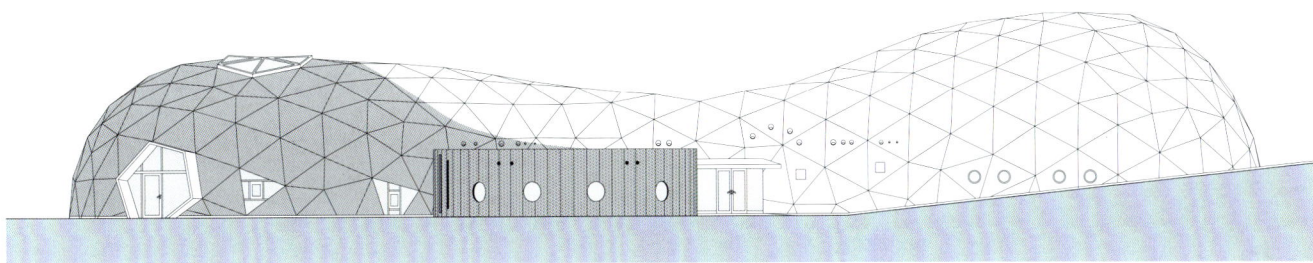

西立面図 S=1/200

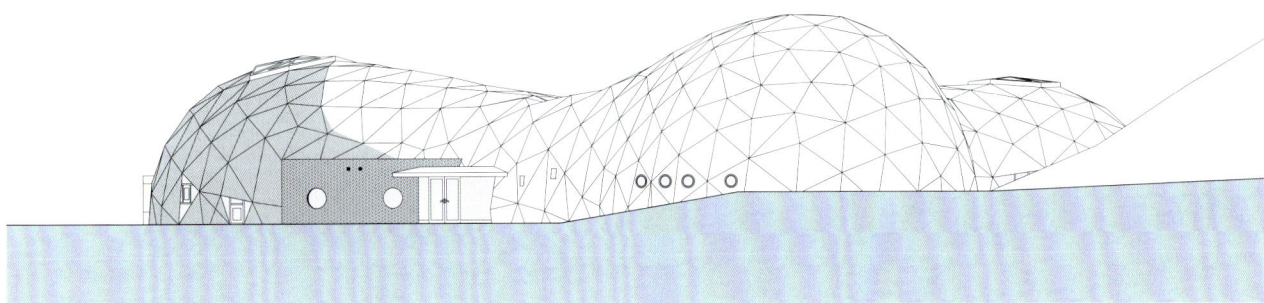

南立面図 S=1/200

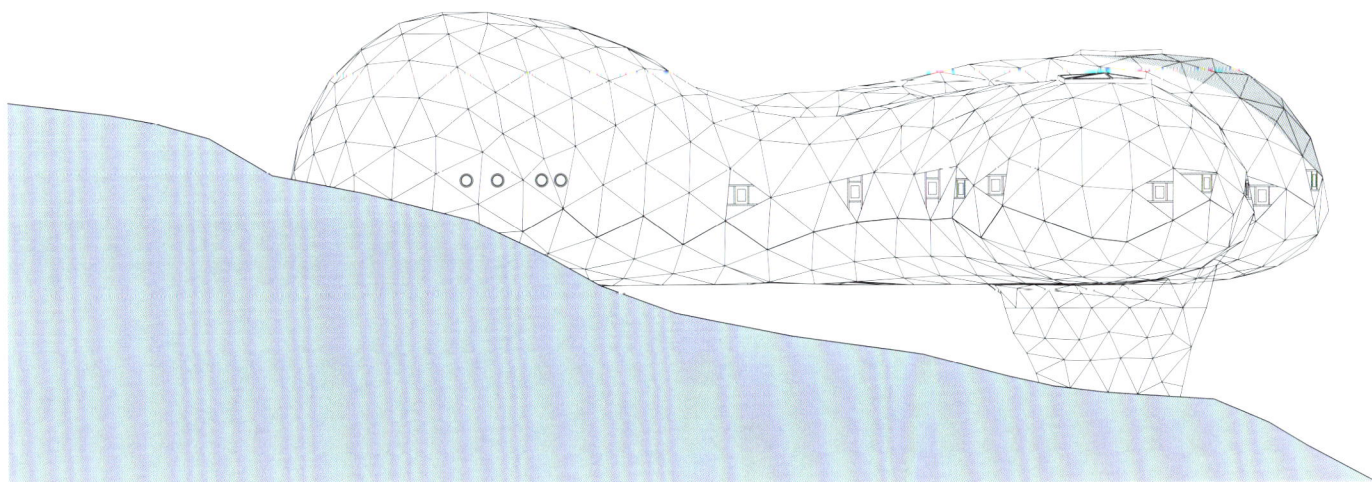

東立面図 S=1/200

1 わんぱく広場(図書・情報コーナー)
2 玄関
3 シアター
4 地球工房(ワークショップルーム)
5 倉庫
6 機械室
7 防火水槽
8 雨水貯留槽
9 中庭

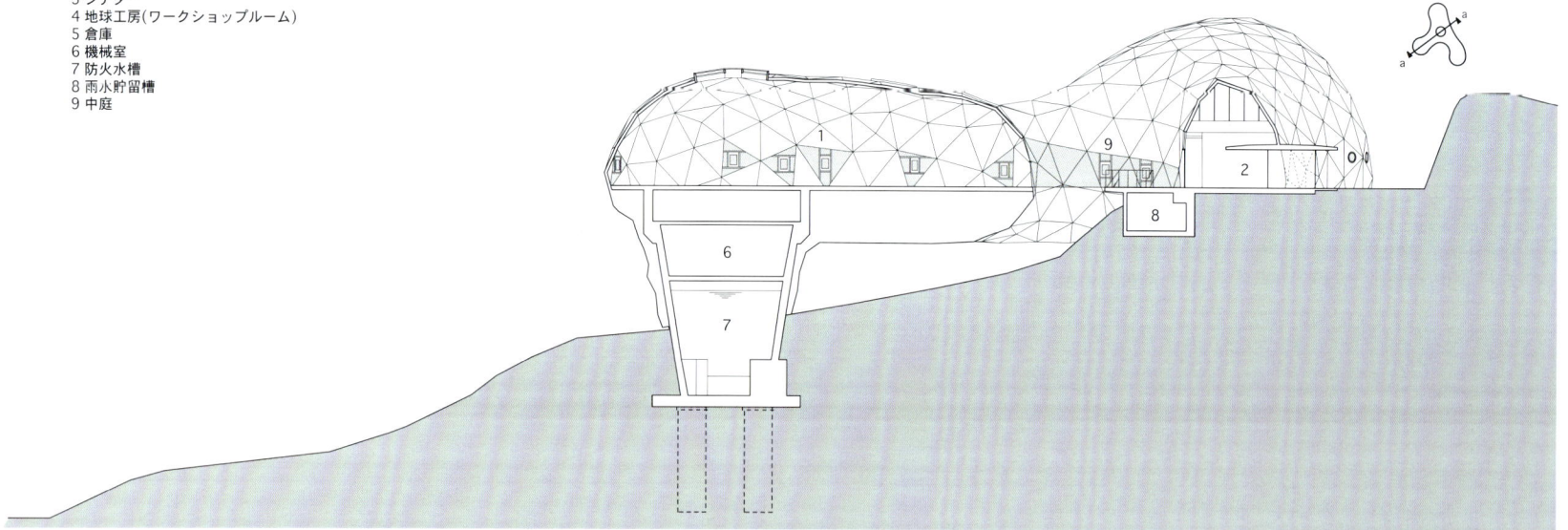

a-a断面図 S=1/200

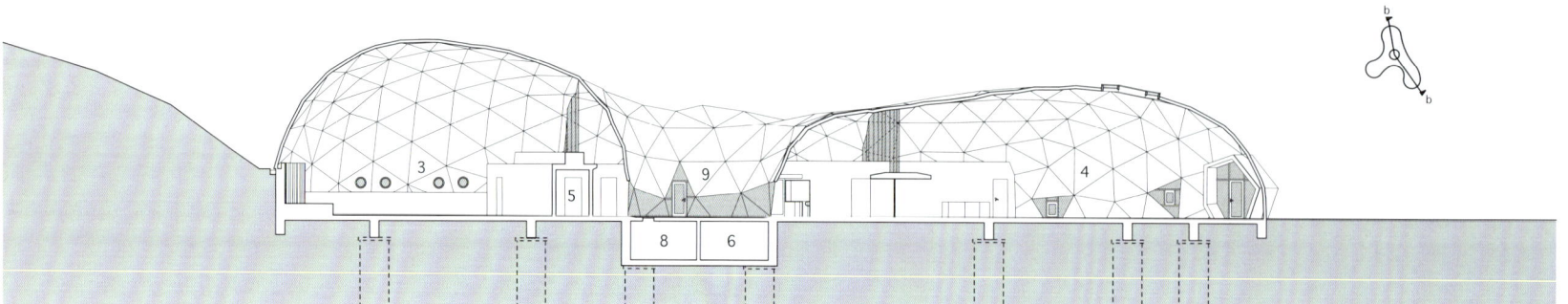

b-b断面図 S=1/200

　　业主要求建筑用作游客中心，使游客和兵库县居民都乐于拜访，以此提高人们对全球环境问题的关注度，让他们从各方面更好地领悟环保的意义。建筑师想到创造一个新的环境建筑空间，来分享环境和自然的接触点，其中的关键词是：流通。

　　这个建筑位于北面山上的陡坡上。在综合了业主关于设计方案的所有功能要求后，建筑分为三部分，其功能和结构的安排则根据必要的区域和体积来考虑。两个功能区被安排在工地上突出的两块平行的平地上，那里连通原来的城镇小路，另一个区域也悬浮在陡坡上，与其他两个呈同一水平线。这是为了确保每一块有限的平地都能得到最大的利用，并尽可能地保持自然景观的原貌，以及减少建筑对周围自然环境的破坏。这种形式的建筑体现了无论从平面上，还是从截面上来看，它都是能连接起三个功能区的合理的结构。

The Blakehurst Club

Architect: Tony Owen
Firm: Tony Owen Partners
Location: Sydney

The Kyle Bay Bowling Club is located on a spectacular waterfront location in Blakehurst in Sydney's southern suburbs. This design featured a unique diamond shaped parabolic roof. Despite the roof design the upper level of the club had low ceilings and did not take advantage of the great views and location.

The designer looked several options including raising the roof with a paraboloid design to create a more dramatic space. Ultimately the designer reinstated the roof pitch. Internal the designer took the opportunity to create a space which maximizes the potential of the roof form and the location. By locating all of the air conditioning in a perimeter plenum the designer was able to maximize the ceiling volume to create a light and airy space which connected with the outdoors. The distinctive ceiling is based on the wing structure of the wetland insects found on the site.

The designer made extensive use of 3-Dimensional digital modeling software to explore different options for the roof geometry. Eventually chose a branching "L-system" geometry for the coffered roof. This geometry reflects light throughout different times of the day to create an effect that echoes the play of light on the water outside.

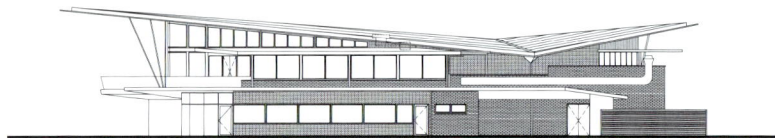

01 EAST ELEVATION

02 NORTH ELEVATION

01 WEST ELEVATION

02 SOUTH ELEVATION

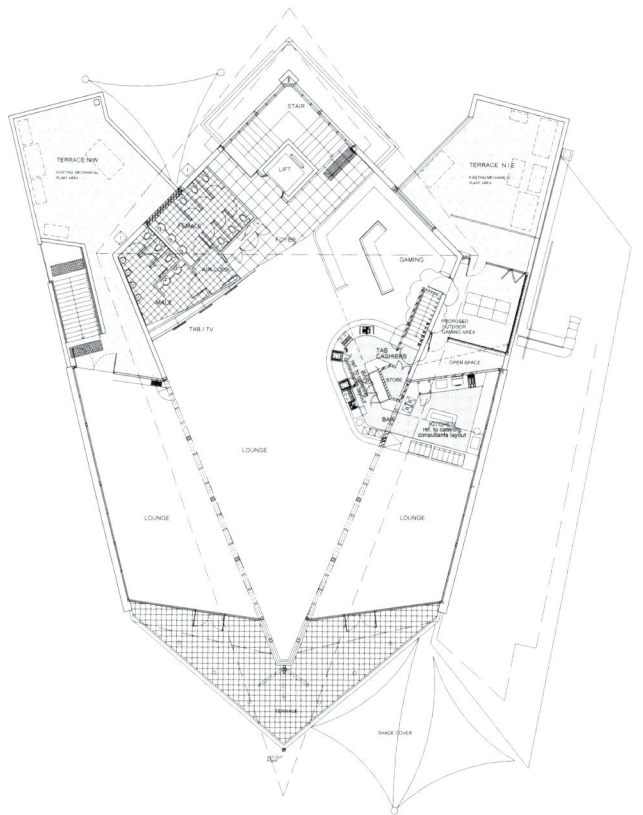

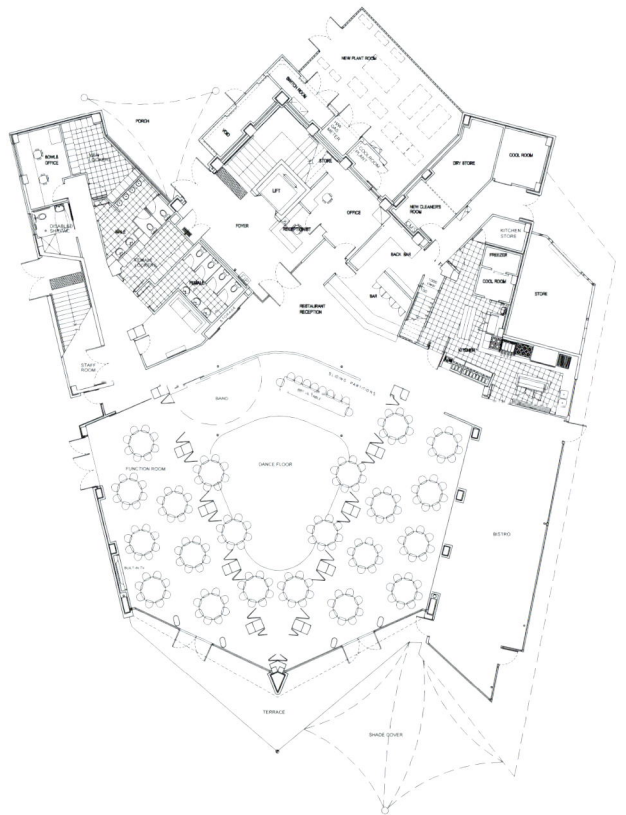

这个建筑坐落在悉尼南部郊区的布列克赫斯特，一个景色壮观的海滨旁边。这个建筑独特的设计是它有一个钻石形的抛物线形屋顶。原有屋顶的设计尽管提升了俱乐部的档次，但它过于低矮而未能充分利用这个风景优美的绝佳地点。

设计师寻求了几种方法来提升屋顶高度，其中包括用一个抛物线的设计。最后设计师修复了屋顶坡度。在内部则利用各种机会创造空间，使屋顶的形状和位置实现的可能性达到最大化。通过周边送气通风室对空气调节的定位，使天花板的体积达到最大化，并建立了一个与户外连接的既通风又明亮的空间。这个与众不同的屋顶的设计是从湿地的昆虫的翅膀结构得到的灵感。

设计师广泛运用三维空间的数字建模来探索屋顶几何形状的不同选择。最终选择了几何学的"L系统"来建造方格形的屋顶，这样建筑的光线便能倒影在户外的水面上。

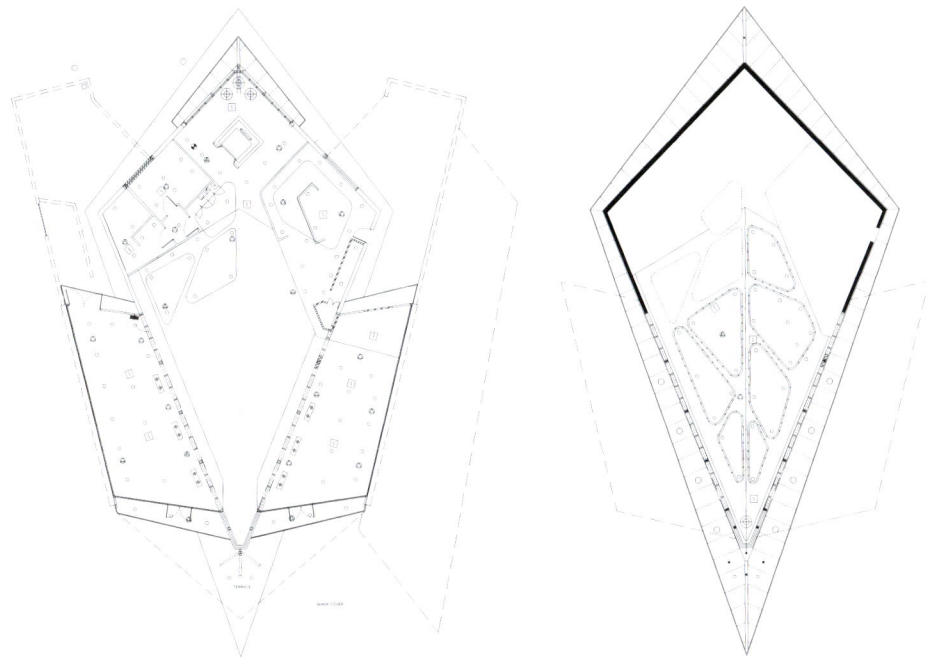

Sluňákov - Center for Ecological Activities

Architect: Roman Brychta, Adam Halír, Petr Lešek, Ondrej Hofmeister
Firm: Projektil Architekti
Collaborators: Katarína Jägrová, Katerina Horáková, Lenka Slívová
Location: Olomouc, Czech Republic
Area: 1586m²
Photographer: Andrea Lhotáková

The Slunakov – Center for ecological activities of the city of Olomouc has been designed as part of a project entitled "Slunakov, facilities for ecological activities – educational biocenter".

During the process of construct, the designers build a kind of new ccological architecture. The building has been designed as a curved inhabitable land wave that fluently blends into the surrounding terrain and symmetrical following the exact North-South axes. All materials used are traditional and have been chosen due to their environmental friendliness. The facades are covered by wood, glass, concrete and stone (stacked). The interior is completed using mainly wood, glass and brick walls plastered or, in the case of unfired brick left uncovered. Fired brick or reinforced concrete is used in the supporting structure for the technical rooms and wet activity areas. Most floors are covered by wooden planks and those in wet activity areas or in technical areas are covered by seamless floor.

The measures, such as a heating and ventilation using heat recovery to heat the building, solar collectors for hot water preparation and for support of space heating and earth heat exchanger, will also be used for educational and demonstration purposes. Ventilation and warm air heating are ensured by fresh air and by warm air circulation ventilation with heat recovery from outgoing air.

entrance to the park and parking   building   educational biocentre   town

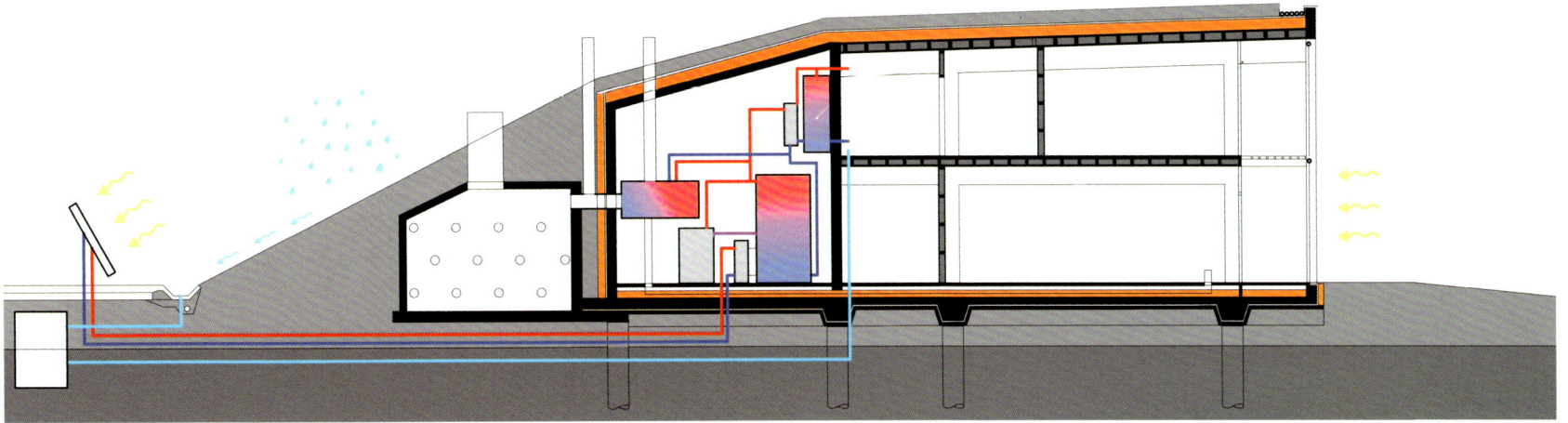

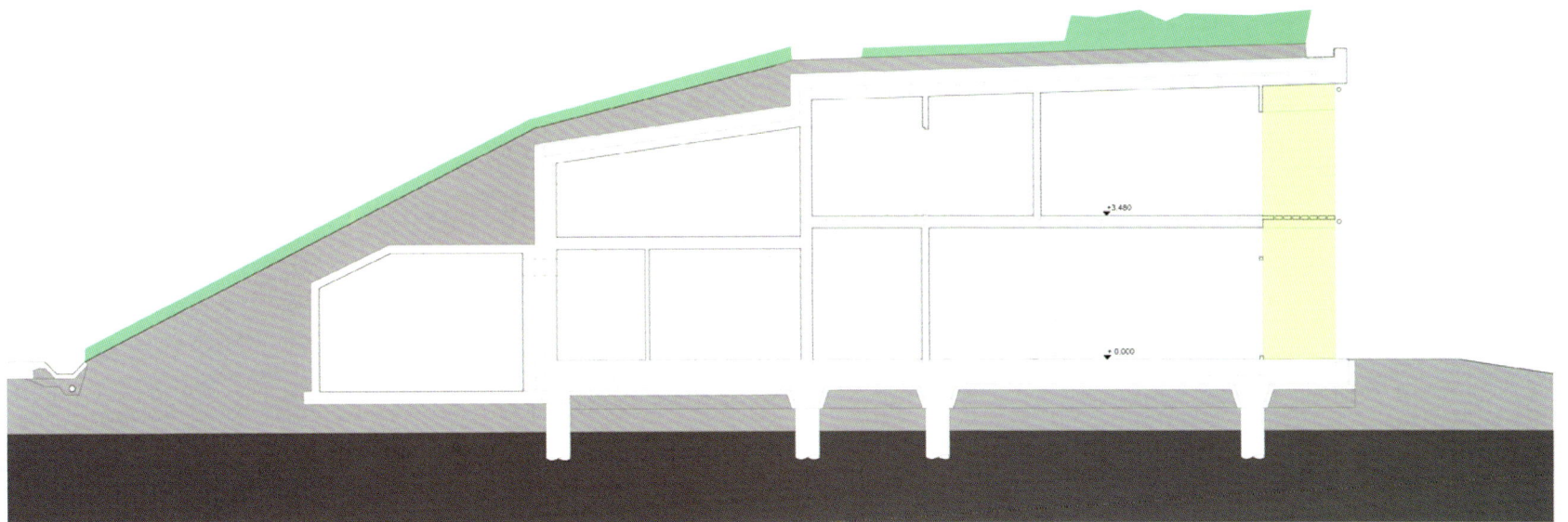

B - B

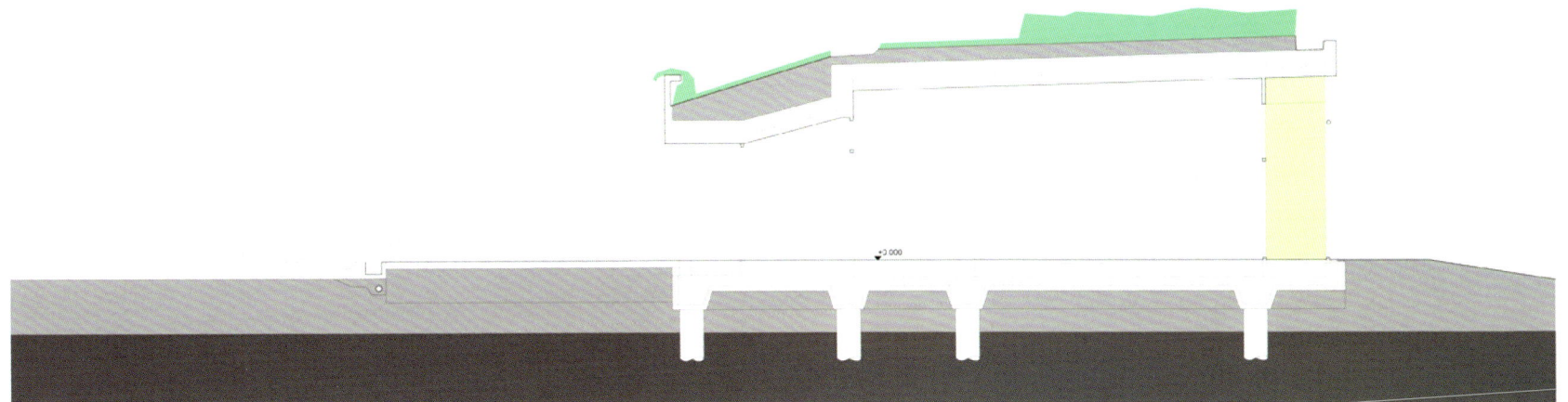

A - A

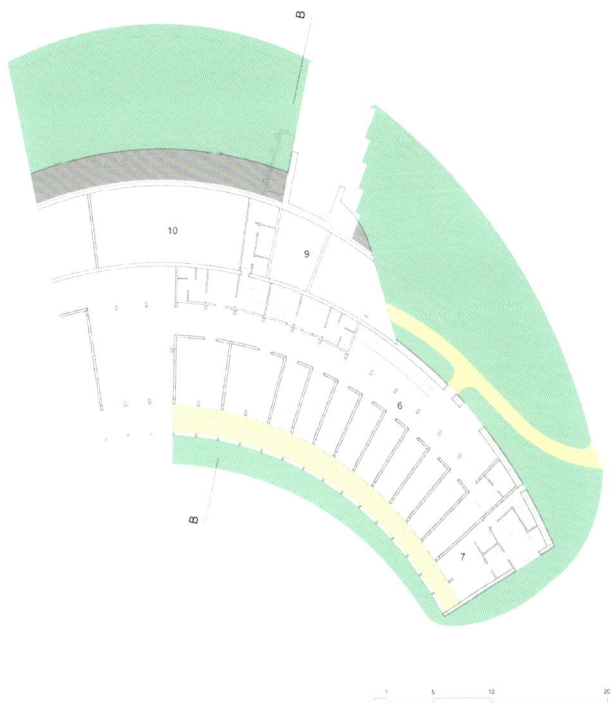

6 - hostel rooms, 7 - caretaker's flat, 9 - technology, 10 - storerooms

1 - administration, 2 - entry hall, 3 - multifunctional hall, 4 - canteen, 5 - lecture rooms, 6 - hostel rooms, 7 - caretaker's flat, 8 - kitchen, 9 - technology

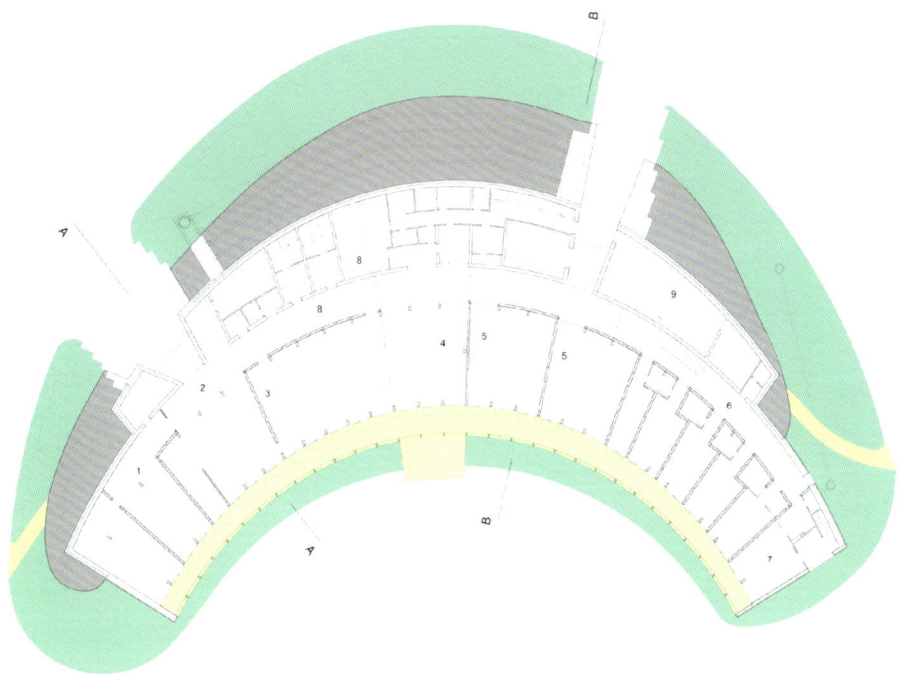

šatna

Slunakov——奥洛穆茨生态活动中心，是"Slunakov 生态环境活动和教育"的项目之一。

构建项目过程中，创造了新形式的生态建筑，不仅融入了周围环境，而且参照南北轴线成对称结构。所有使用的材料都基于环保理念。外部主要运用木材、玻璃、水泥、石头。内部主要利用木材、玻璃和砖墙贴。钢筋混凝土被用作支撑结构。大多数地面覆盖木板，形成无缝地板。节能措施，如暖气和通风热回收利用，太阳能和地面加热换热器，也起到教育和示范的作用。通过通风和加热暖空气，确保空气新鲜和空气循环通风与散热。

Arhus Gymnastics and Motor Skills Hall

Architect: C. F. Møller Architects
Firm: C. F. Møller Architects
Location: Hvidkildevej, Århus
Area: 1600m²
Photographer: Poul Nyholm, Julian Weyer

The Motor Skills Hall is an extension of the Aarhus Gymnastics and Trampoline Hall. The idea of the augment has been to design a hall which motivates the 3-10-year-old children to practise movement. A sculpturally-formed climbing frame, called the seashell, extends across the hall, stretching right up under the ceiling, there is a look-out tower at a height of eight metres providing a view of Aarhus Bay and Mols Mountains. The hall also invites movement through its many footbridges and platforms located along the length of the hall, and flexibly furnished zones called kangaroo land with trampolines and jumping equipment, and a so-called monkey land with ropes and jungle-like density. The unusual spaces which arise through the extension have also been put to active use; the slanting roof surface of the existing hall has for example been transformed into an indoor climbing wall. From the outside, the hall has the appearance of a raw concrete shell, with large, slanting windows and stairways which lend transparency and dynamics to the facade.

运动技能馆是对奥尔胡斯体操馆和蹦床馆的扩建工程。扩建的主要目的是为3至10岁的儿童设计一个训练场所。精心设计出来的攀登架，名字叫做贝壳。"贝壳"沿着场馆在天花板下延伸开来，结合旁边八米高的瞭望台，看起来就像奥尔胡斯湾和莫尔斯山。场馆内也可以让人沿着场馆的人行天桥和平台穿梭运动。场馆内还有一个经过灵活布置的区域叫做袋鼠园，里面有蹦床和蹦跳设备，还有一个猴子山，里面有许多绳索和假丛林。扩建后的一些特别的空间已经投入使用。譬如，原来倾斜的屋顶已经被改造成为室内攀岩墙。从外面看，场馆像由混凝土制成的贝壳。大而倾斜的窗户和楼梯更让它从正面看起来明亮而具有动感。

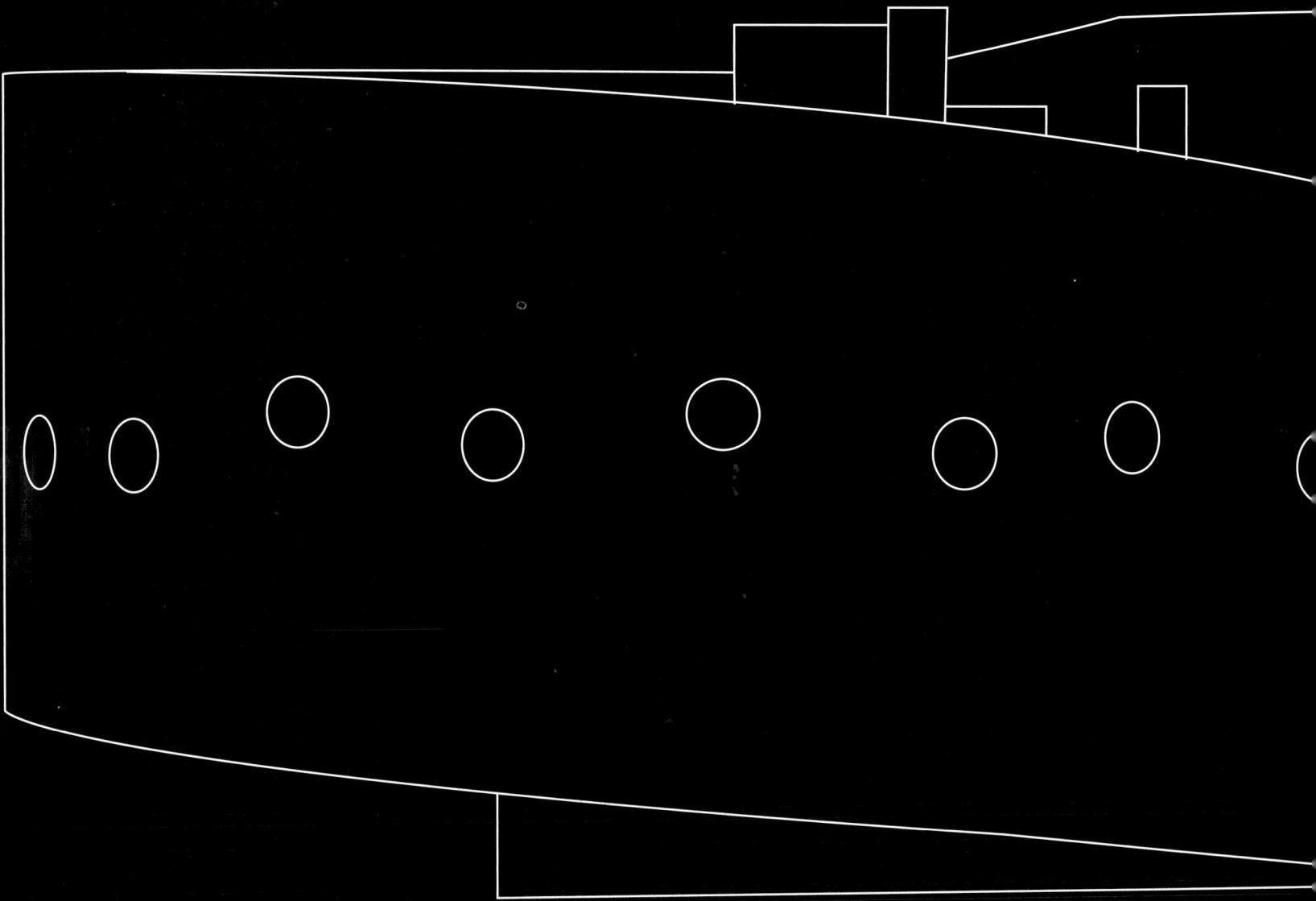

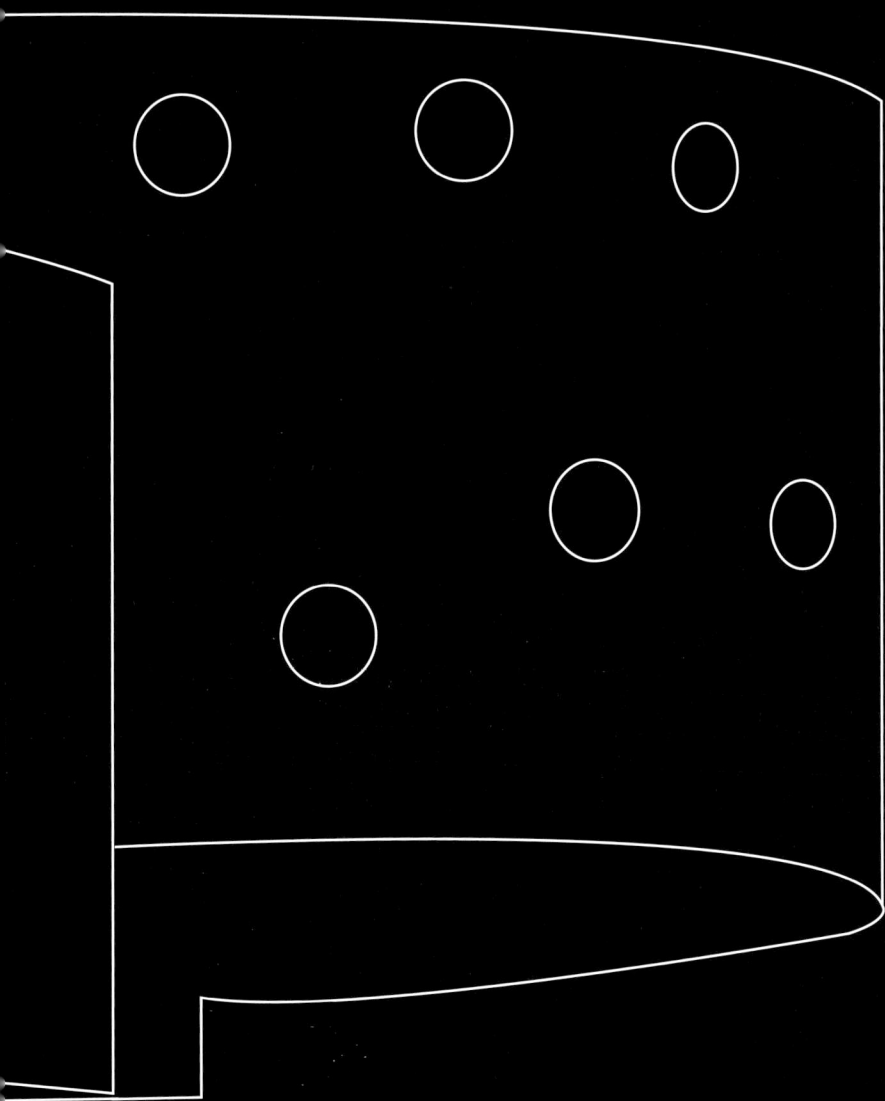

Other

其他

Maritime Center Vellamo

Architect: Ilmari Lahdelma
Firm: Architects Lahdelma & Mahlamäki
Location: Tornatorintie 99, 48100 Kotka, Finland
Area: Floor area 14601m², Gross area 14366m² and Volume 118039m³
Photographer: Jussi Tiainen

The figure of the Maritime Center guides travellers from the heart of the city out to the north-eastern end of the City Terminal and into a harbour of culture. The Maritime Center is the hallmark of the city's cultural profile. The Maritime Center is a notable addition to landmarks associated with Kotka both nationally and internationally.

Situated at the end of the planned culture harbour, the roof of the Maritime Center, the crown of the cityscape, forms a square which will play host to a wide array of different events. A sloped floor with a staggered performance space, an incline culminating in a steel-glass covering and the square's position high above the surrounding city together create a unique space suitable for all manner of different activities. The outdoor exhibition space and the main entrance, opening out towards both the city center and the quayside, form a focal point in the ground-level cityscape.

The interior of the Maritime Center is characterized by the application of timeless architectural concepts. With the help of natural light, spatial layout and the choice of materials, individual spaces come together to form shifting, adaptable, dynamic spatial units.

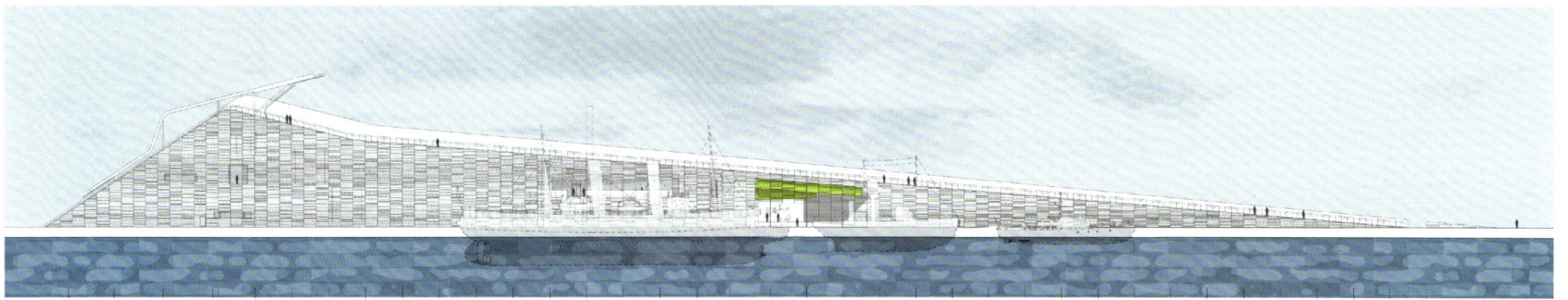

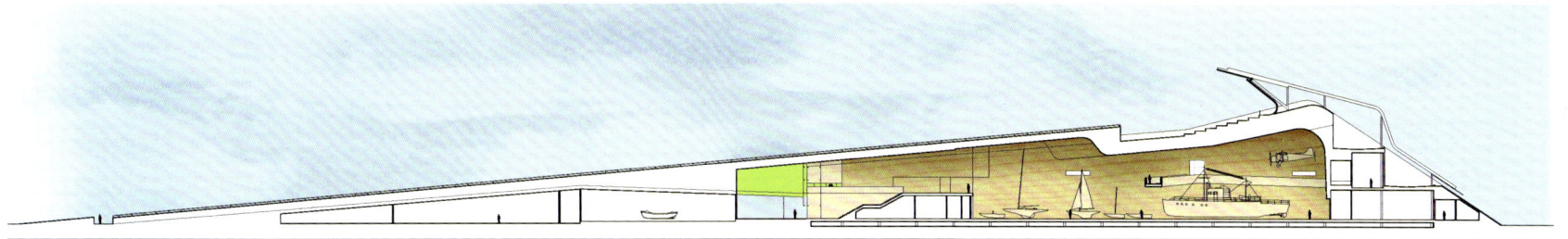

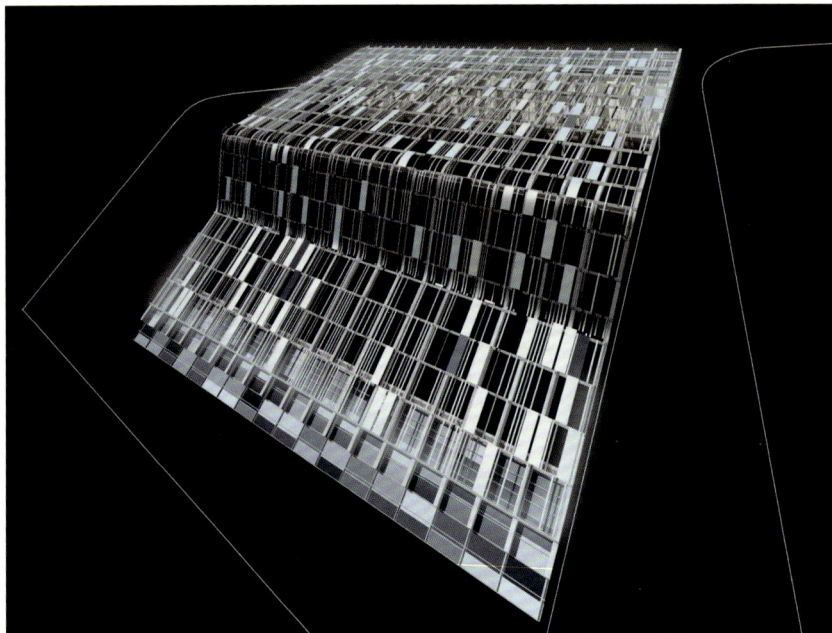

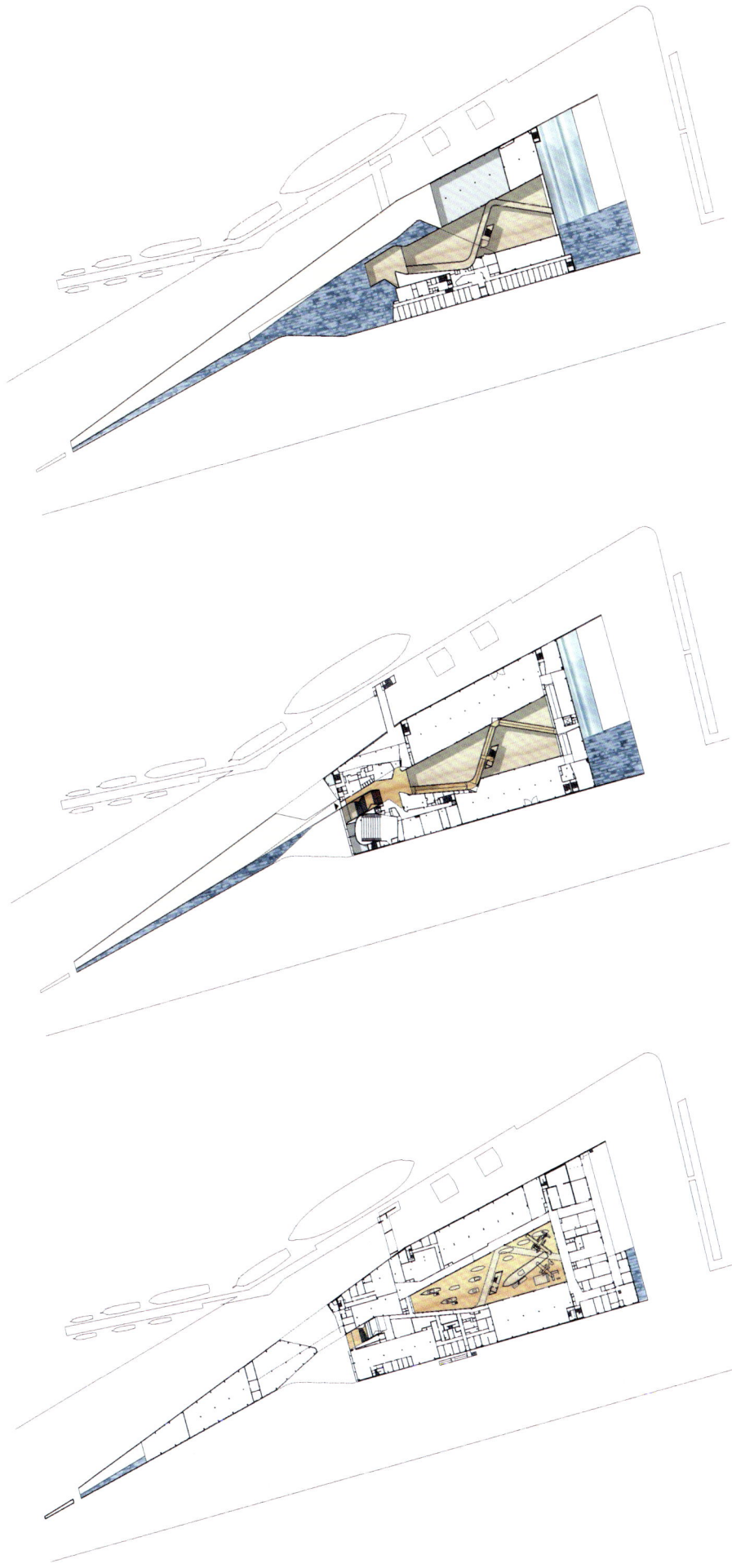

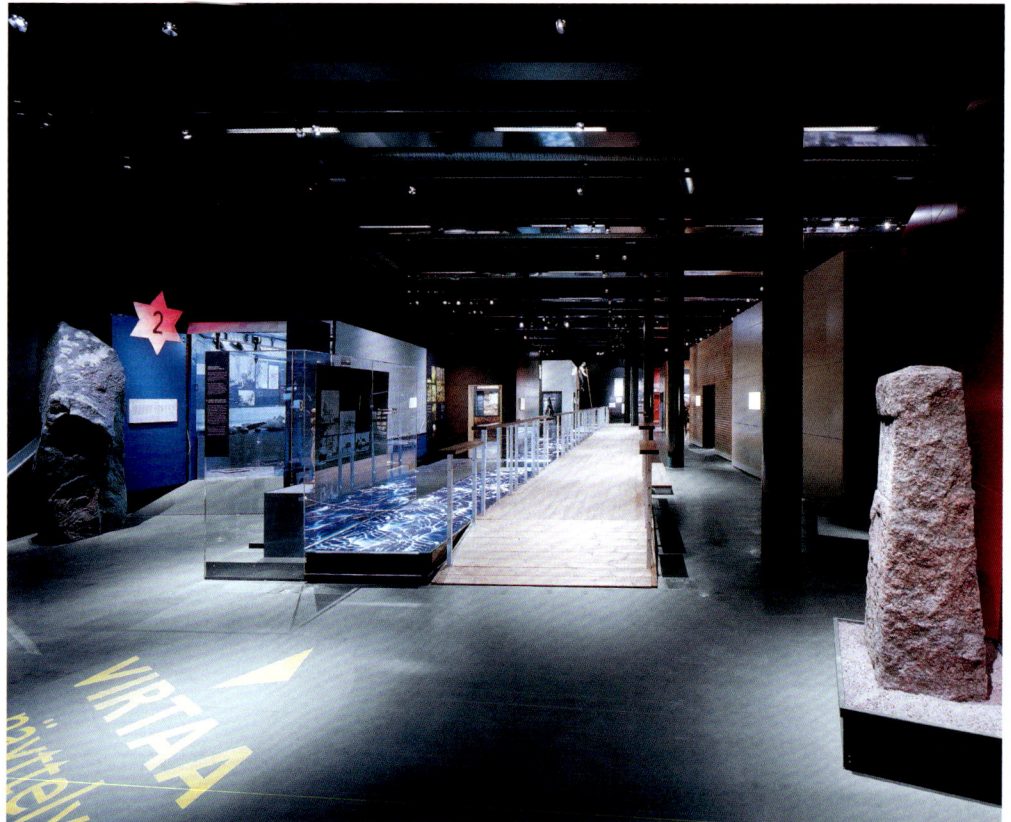

　　航运中心作为地标指引着游客从市中心地带来到东北部，一个文化的港湾。航运中心是这座城市文化形象的标志，也为科特卡在国内、国际上增添了知名度。

　　位于文化港湾的边界处，航运中心的屋顶看起来像是给这座城市戴了一顶皇冠。这里也是举办各种活动的广场。表演台由倾斜的格子地板覆盖，钢玻璃表面慢慢向上倾斜，广场海拔高于市区海拔也正好为各种各样的活动提供了表演场所。室外展览区和主入口面朝市中心和码头，在这座城市的地平面上形成了焦点。

　　航空中心的室内设计特点是对古今建筑概念的运用。自然光线、空间布局和材料的选择以及个人空间的设计相协调，形成了一个多姿多彩、变化万千、动态的建筑整体。

Looptecture Fukura

Architect: Endo Shuhei
Firm: Endo Shuhei Architect Institute
Location: Minamiawaji-city, Hyogo-pref. Japan
Area: 8502m²
Photographer: Yoshiharu Matsumura
Material: Steel

The function of this architecture is security and controlling all the floodgates located at port of Fukura, enlighten dangerous of the tsunami for tourists, and use as a place of refuge in case of the tsunami warning. For these reasons, ensure to keep the spaces of necessary and viewpoint for watching all over the port, also rational shape and structure to against of tsunami and the drift came after disaster are necessary.

Main floor is placed higher level than assumed height of tsunami and opening the ground level floor allowed when waves passing through. Make the outside wall curved as efficient form to disperse stress. This form consists of 7.3m width belt (curved wall); continuous arcs are constructer and crossing by 6 different center of circle. Consequently, these arcs are closing at the same point both start and finish.

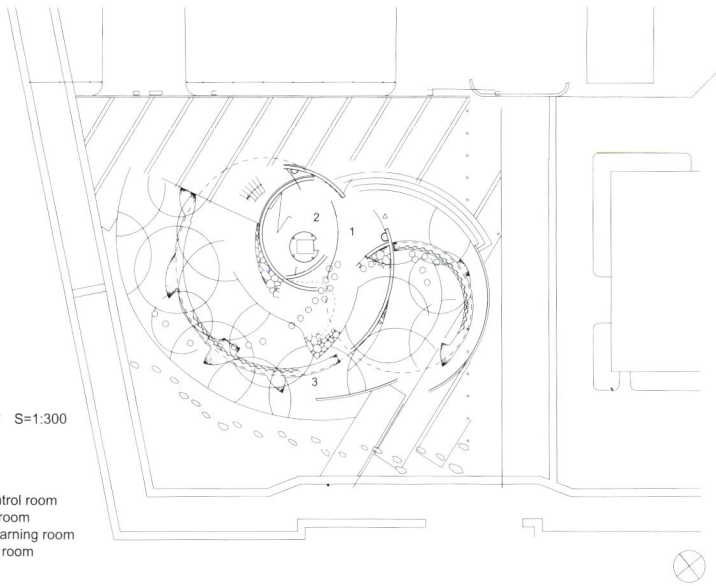

Looptecture F   S=1:300

1. Entrance
2. Hall
3. Piloti
4. Centre control room
5. Exhibition room
6. Disaster learning room
7. Machinery room
8. Lavatory
9. Terrace

Site Plan and 1F Plan   S=1:300

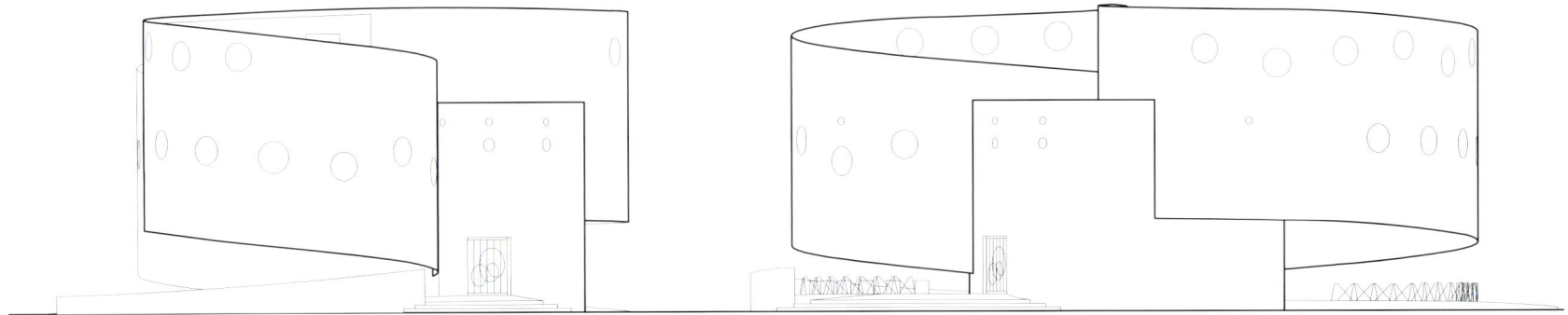

East

North

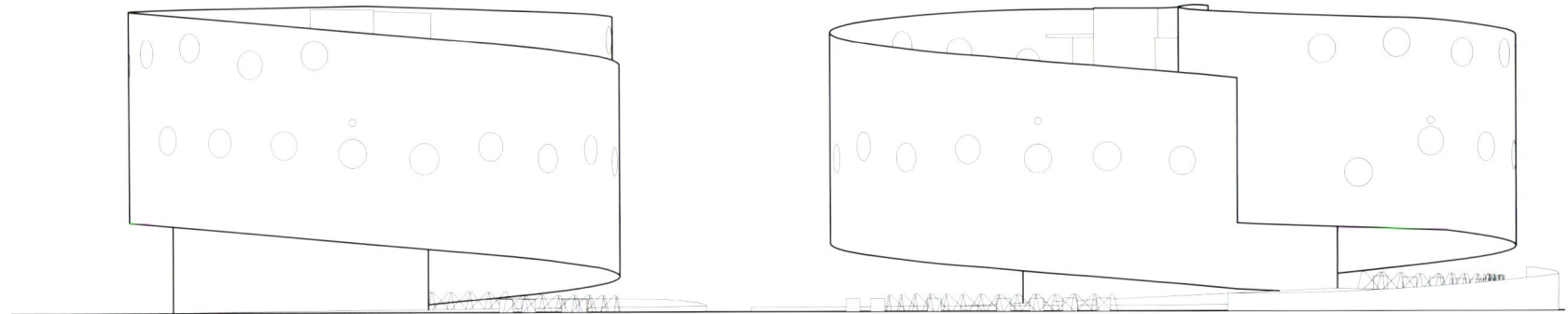

West

South

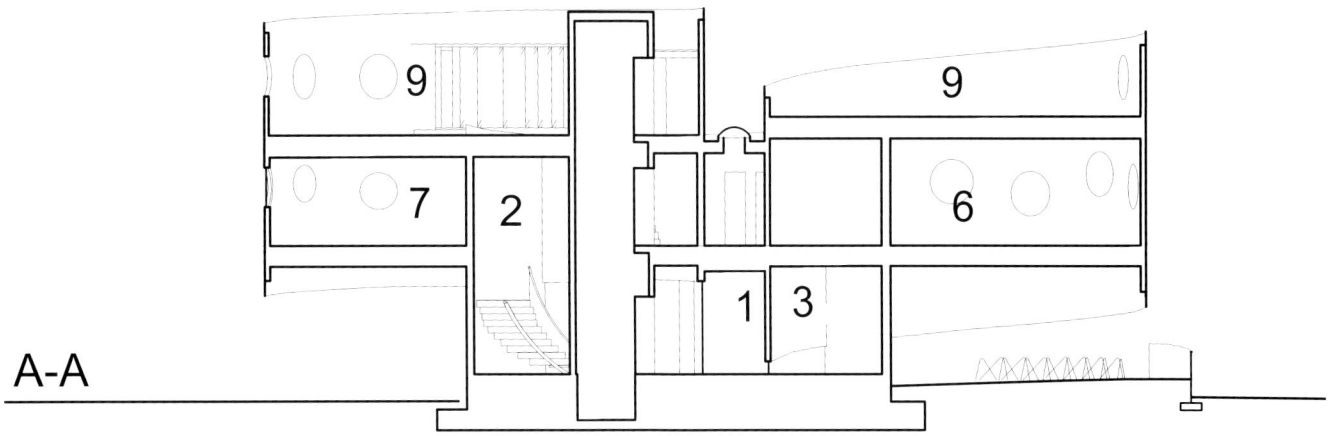

A-A

Looptecture F Sections   S=1:300

1. Entrance
2. Hall
3. Piloti
4. Centre control room
5. Exhibition room
6. Disaster learning room
7. Machinery room
8. Lavatory
9. Terrace

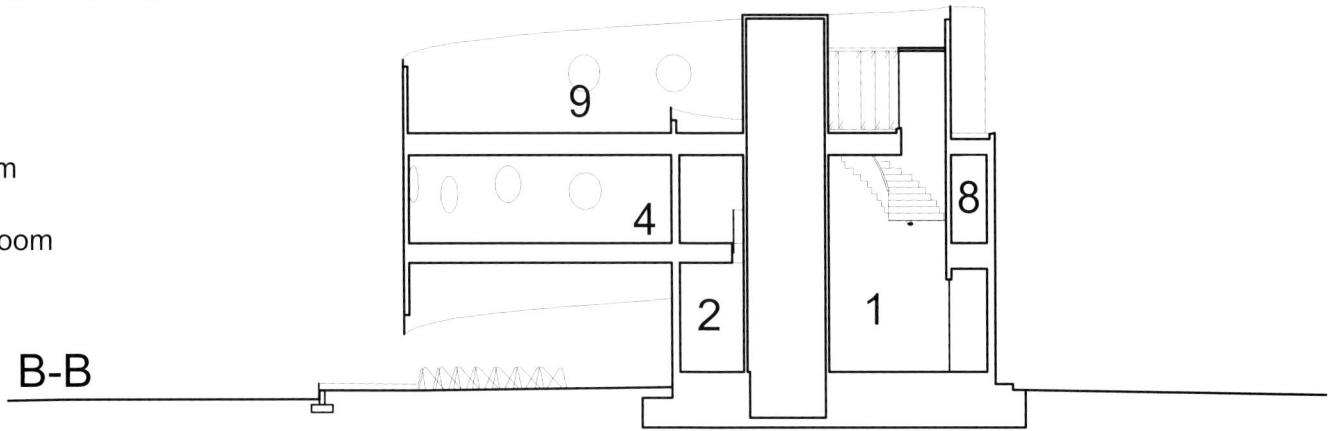

B-B

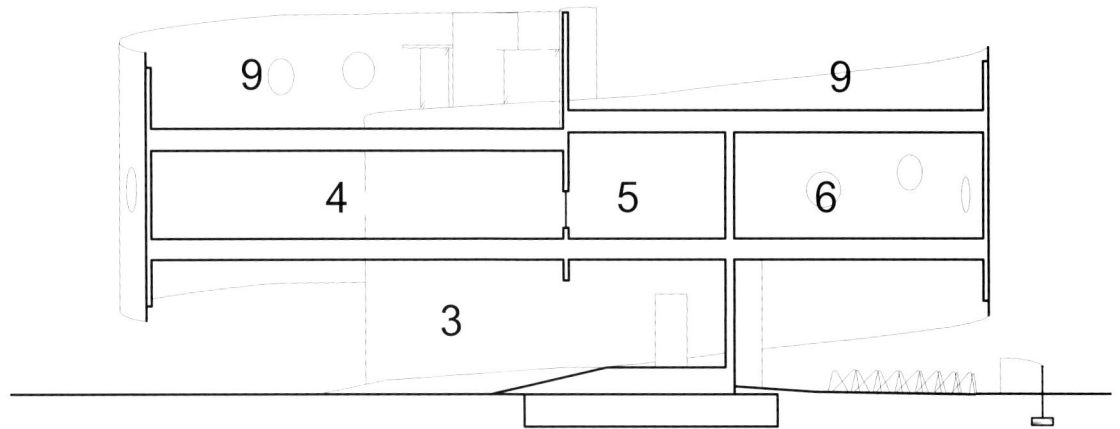

C-C

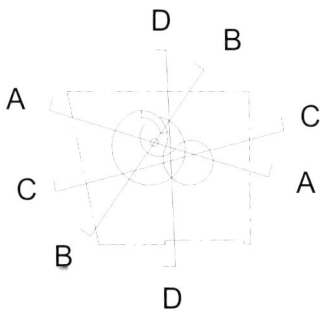

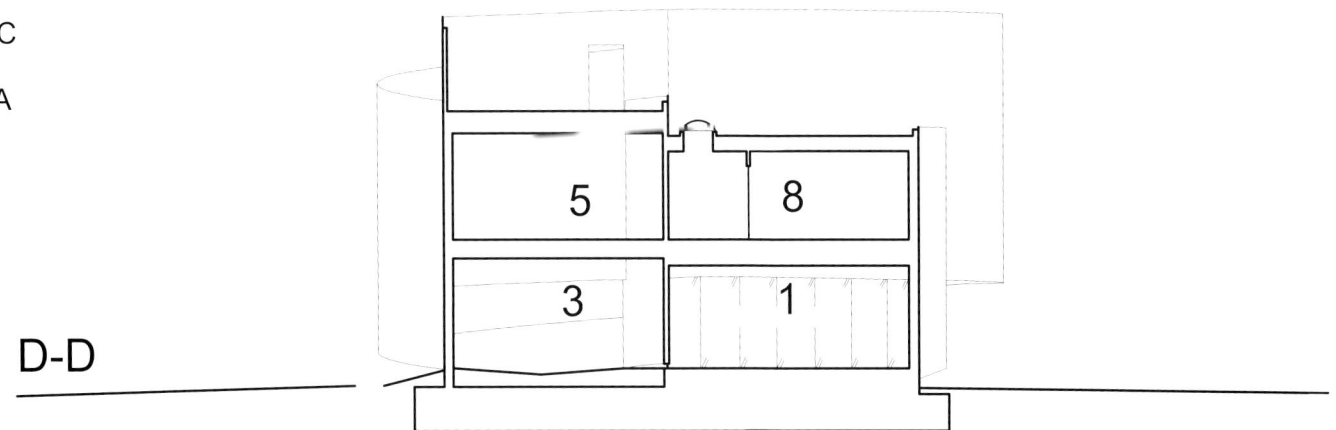

D-D

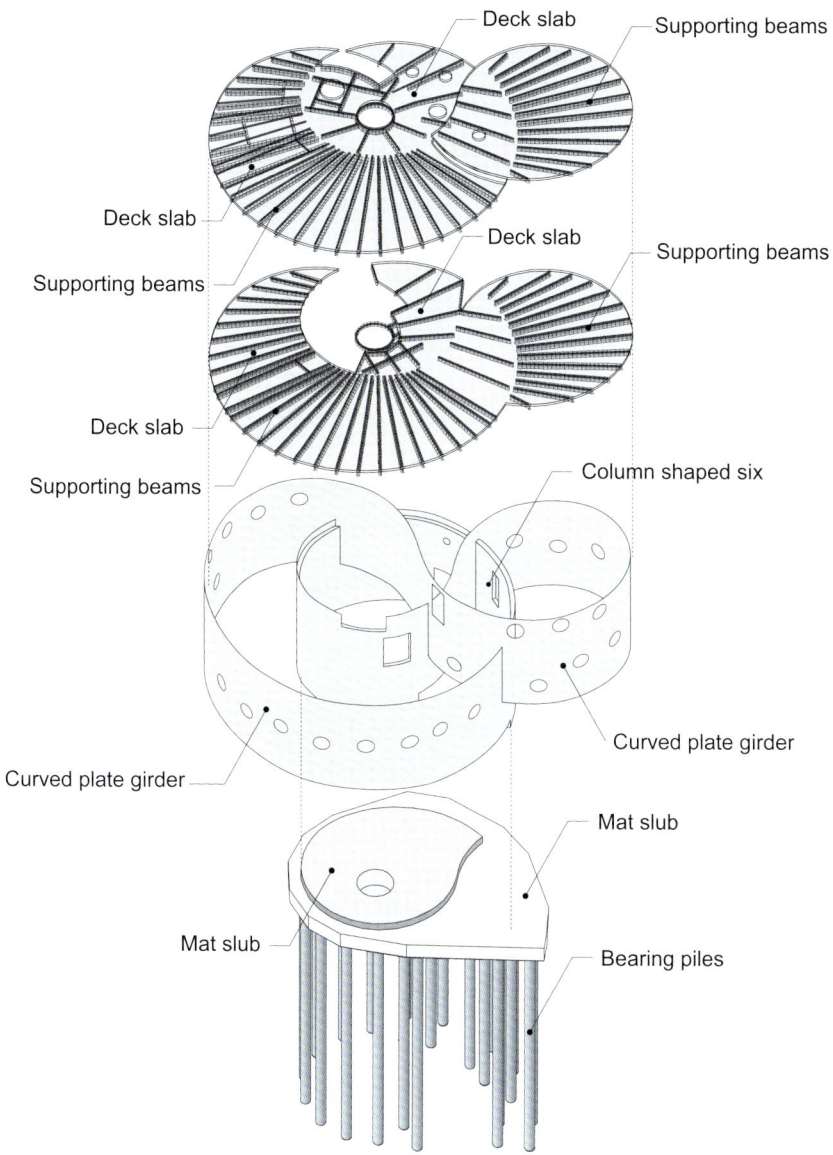

Deck slab
Supporting beams
Deck slab
Supporting beams
Deck slab
Supporting beams
Deck slab
Column shaped six
Supporting beams
Curved plate girder
Curved plate girder
Mat slub
Mat slub
Bearing piles

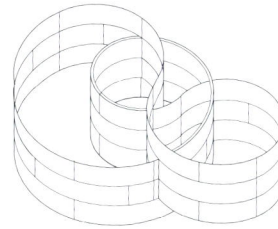

Structure Axonometric

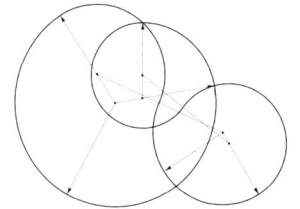

Linear diagram - Integration of six circular arcs -

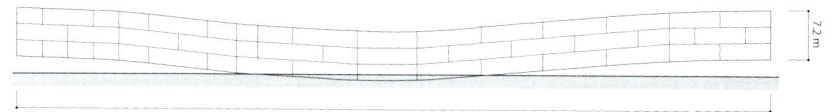

Approx. 120m

7.2 m

Looptecture F   S=1:300

1. Entrance
2. Hall
3. Piloti
4. Centre control room
5. Exhibition room
6. Disaster learning room
7. Machinery room
8. Lavatory
9. Terrace

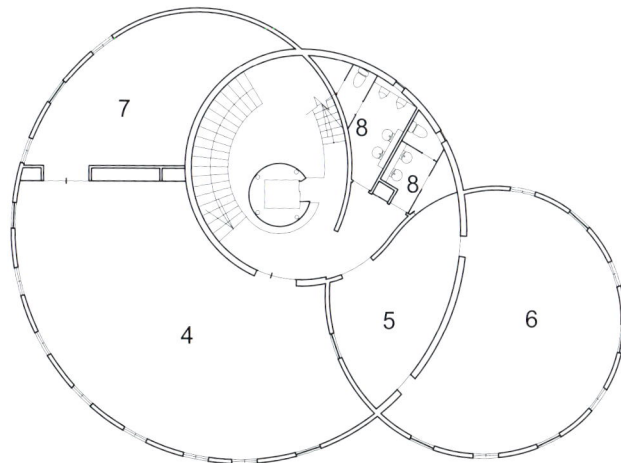

2F Plan

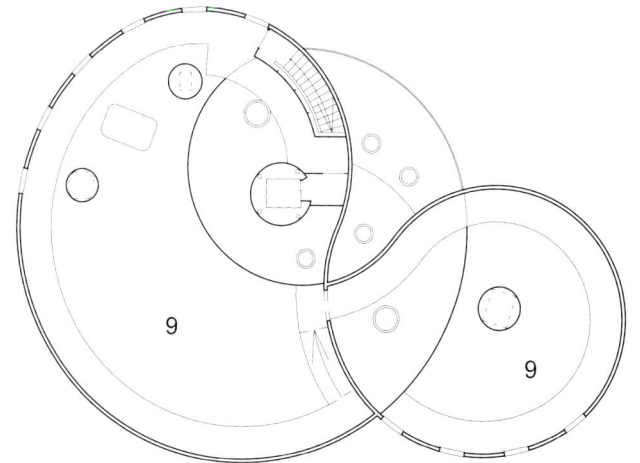

Roof Plan

受海啸危险的启发，这个建筑的功能是控制Fukura港的所有水闸门和保证安全，并用作海啸警报时的避难所。出于这些原因，确保建筑的形状和结构都是为了对抗洪水时提供必需的空间，随时观察所有港口的情况，并能在灾难发生后进行必要的转移。

通过底层的开放设置让海啸时的浪花可以通过，主要的楼层则高于预想的海啸高度。外墙的设计利用有效的曲化来分散压力。这种结构包含7.3米宽的带子和连绵不断的圆弧，还有6个不同的圆圈中心。因此，这些圆弧结束的地方既是它的终点也是起点。

Wooden Boat Center

Firm: Architects Lahdelma & Mahlamäki
Location: Finland
Area: Floor area 1280m², Gross area 1300m²
and Volume 8000m³
Photographer: Jussi Tiainen

Two years ago, the Blue Marlin – a 12mR class racing yacht built in 1938- made its way from Slovenia to Kotka, Finland. This 21-meter classic beauty, even if a little worse for wear, had passed into Finnish hands and taken to Kotka for restoration.

The new premises of the Finnish Wooden Boat Center were designed by the Architect Lahdelma & Mahlamäki Oy. The size of the production hall was determined by the dimensions of the Blue Marlin – self-evidently the first restoration project. The new building was completed on Kotkansaari next to the Maritime Center Vellamo in May 2008. The old halls connected to the new building serve the needs of the shipwrights.

The most distinctive and dominant feature of the boat center is the parable-shaped open-ended dome clad with patinated copper sheets. The 15-meter-high space supported by steel arcs house both the boat-building hall complete with the necessary heating and humidity control systems, and unheated storage rooms on two levels. The dome and the old engineering workshop are interconnected by a wedge-shaped section housing a café, library, a small exhibition room and offices.

两年前，"蓝枪鱼"号——一艘12mR级别赛艇从斯洛文尼亚运到了芬兰的科特卡进行修复。即使这艘游艇外形有一点磨损，但仍然无损它的古典之美。

这个崭新的游船中心是由建筑师Lahdelma & Mahlamäki Oy.设计的。中心的生产车间的大小是由"蓝枪鱼"号的规模决定的，毫无疑问这是它的第一个修复项目。它是在2008年5月在Kotkansaari落成的，旁边紧挨着Vellamo海事中心。旧的工程车间与新的建筑相连，继续为造船工匠服务。

这个游船中心最与众不同的特点是它寓言般开放式的圆屋顶，用能产生光泽的铜片覆盖着，显得古香古色。在15米的高空上，支撑着这两个造船大厅的是一个钢弧形房子，它由必要的供热系统、湿度控制系统，以及上下两层的两个无供暖设施的储藏室组成。连接圆顶和旧工程车间的是一组楔形的房屋，分别是咖啡厅、图书馆、一个小型展览厅和办公室。

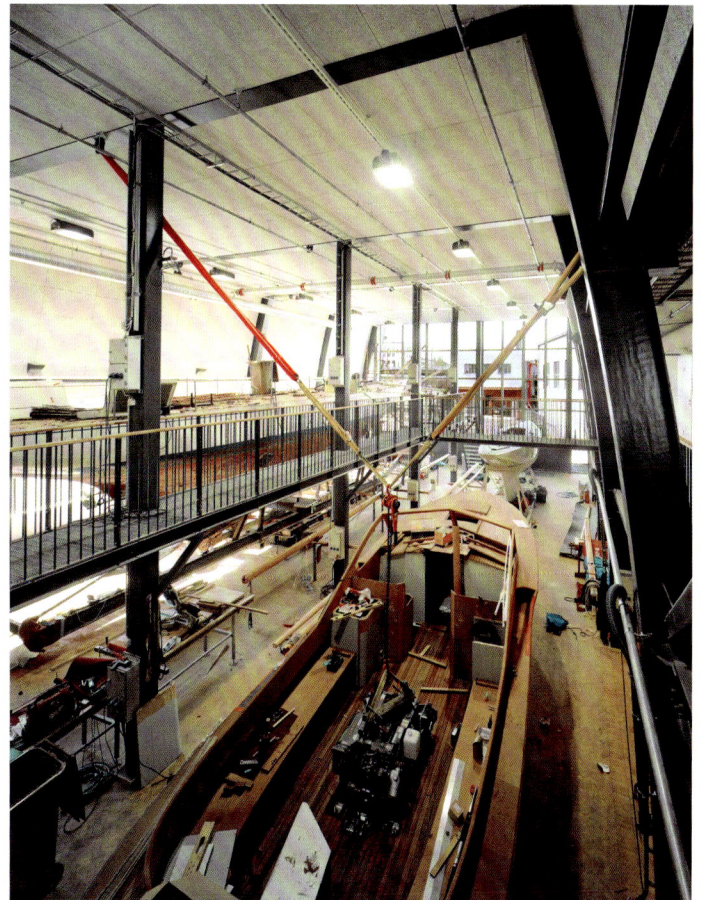

Aurland Lookout

Architect: Todd Saunders & Tommie Wilhelmsen
Firm: Saunders Architecture & Wilhelmsen Architecture
Location: Norway
Area: length: 30m, width: 4m, height: 9m
Photographer: Todd Saunders, Nics Vik & Wisc Vik

- -

Nature first and architecture second was the guiding principal when the designer sat down to design this project. It was immediately obvious that in such beautiful surroundings one must make the least possible encroachment in the existing landscape and terrain. The landscape is so fantastic that it is difficult to improve the place, but at the same time very easy to destroy the atmosphere by inserting too many elements into the site. Even though the designer has chosen an expressive form, the concept is a form of minimalism, in an attempt to conserve and complement the existing nature.

To make the situation even more dramatic it was important to create the experience of leaving the mountainside. The designer wanted people to come out in the air. The construction creates a distinct horizon; a bridge in the open room of this large fjord. It is imperative that the landscape and the vegetation not altered, but are protected so that one can come out from the landscape and experience it from new standpoint.

The designer has managed to behold all of the large pine trees on the site. This allows to create an interaction between the structure and nature. One can walk out onto the air through the treetops, helping dramatizes the experience of nature and the larger landscape room.

Emergency and Infectious Diseases Unit, SUS, Malmo

Architect: C. F. Møller Architects in collaboration with SAMARK Arkitektur & Design
Location: Malmö, Sweden
Area: 26000m²
Photographer: Joergen True

The cylindrical emergency and infectious diseases unit at Malmö University Hospital, Sweden, is designed to minimize the risk of spreading diseases. The distinctive shape also provides a new landmark for the hospital complex. Patients enter the isolation ward via an airlock from the walkway that surrounds the entire building. The exterior lifts are used exclusively by patients of the infectious diseases unit and for hospital waste, while the interior lifts are used to transport staff, supplies and clean materials. Each storey can be divided into sealed-off smaller units in the event of an epidemic. C. F. Møller Architects furthermore designs fixtures for the emergency and infectious diseases unit.

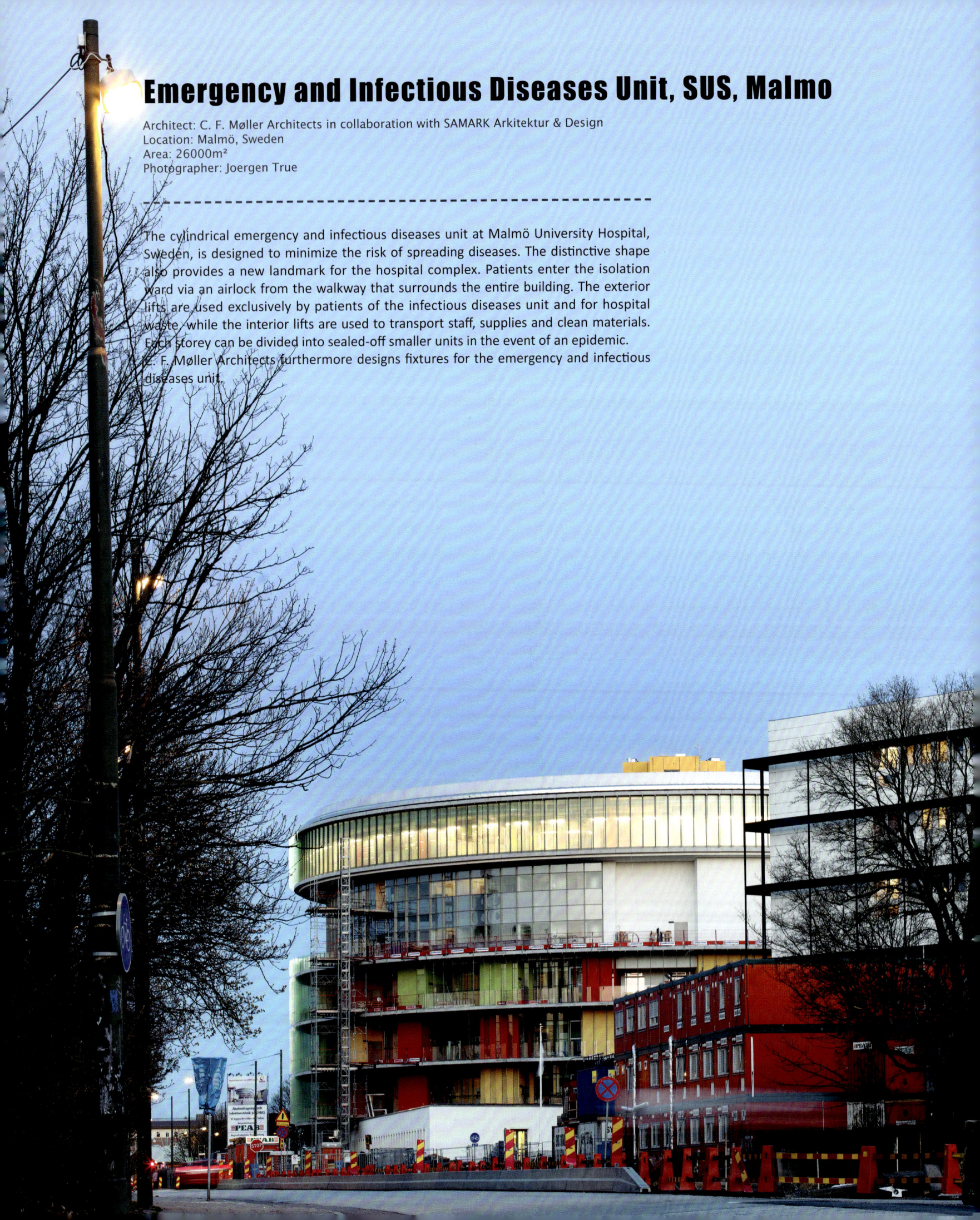

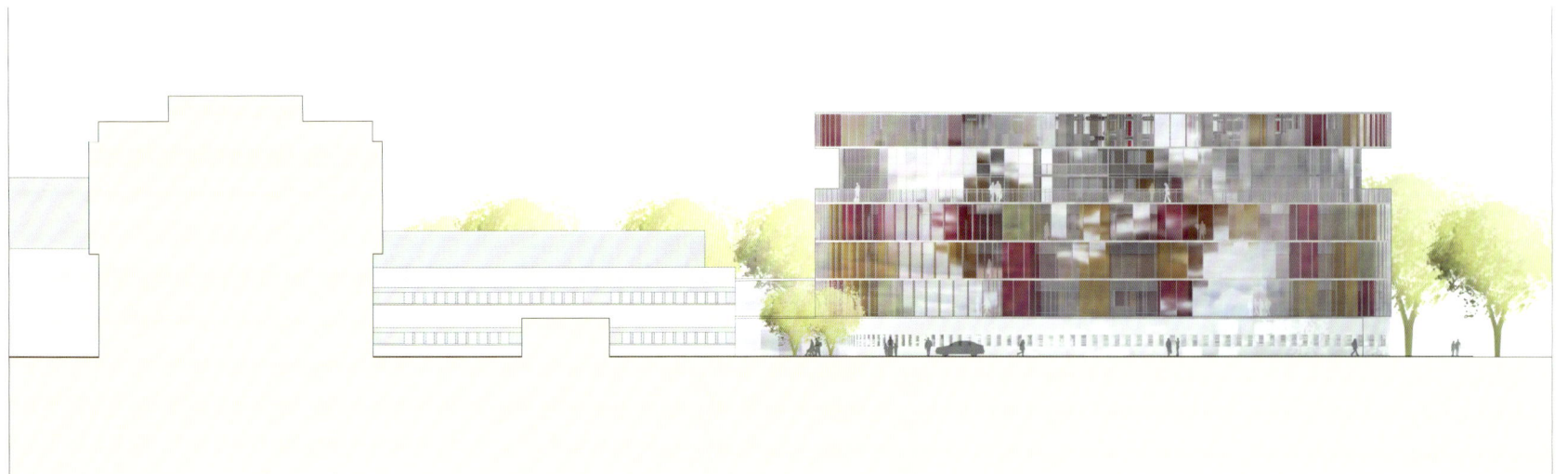

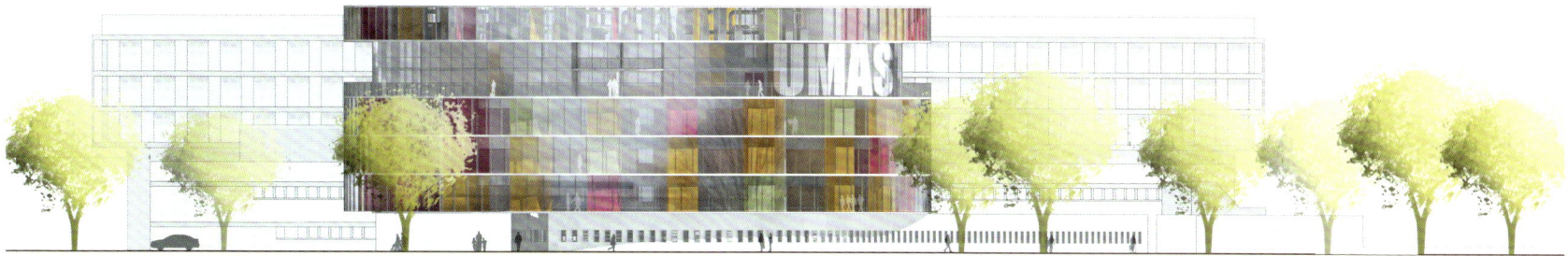

east_facade

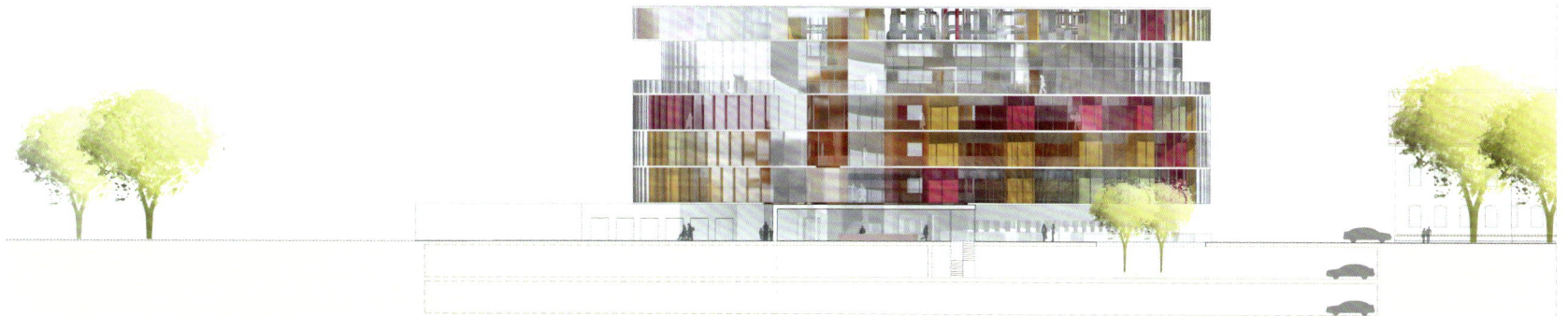

west facade

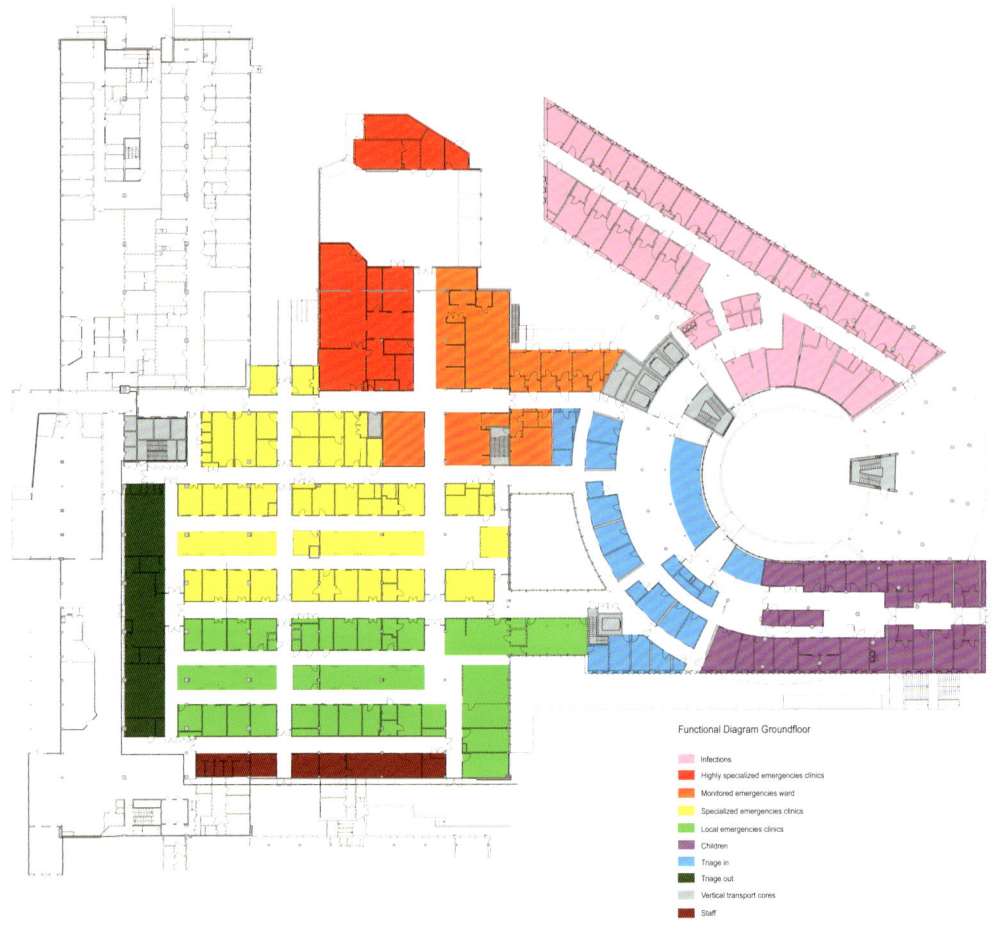

Functional Diagram Groundfloor

- Infections
- Highly specialized emergencies clinics
- Monitored emergencies ward
- Specialized emergencies clinics
- Local emergencies clinics
- Children
- Triage in
- Triage out
- Vertical transport cores
- Staff

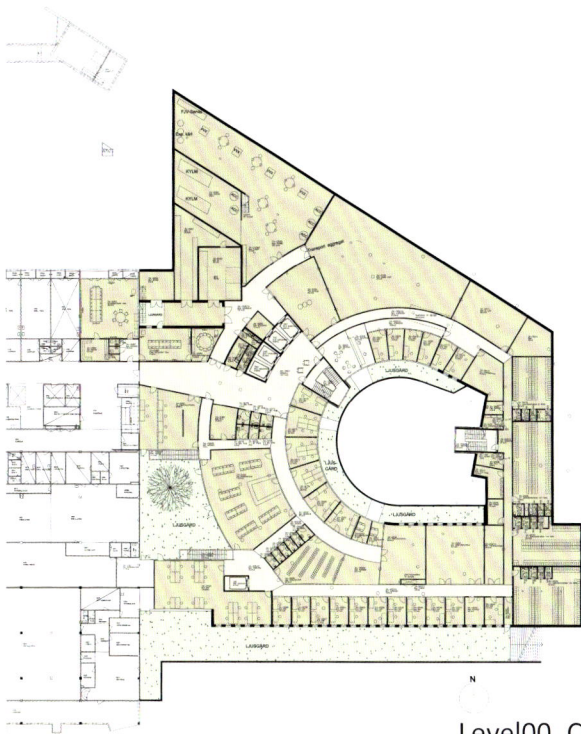

Level00_Clinics support

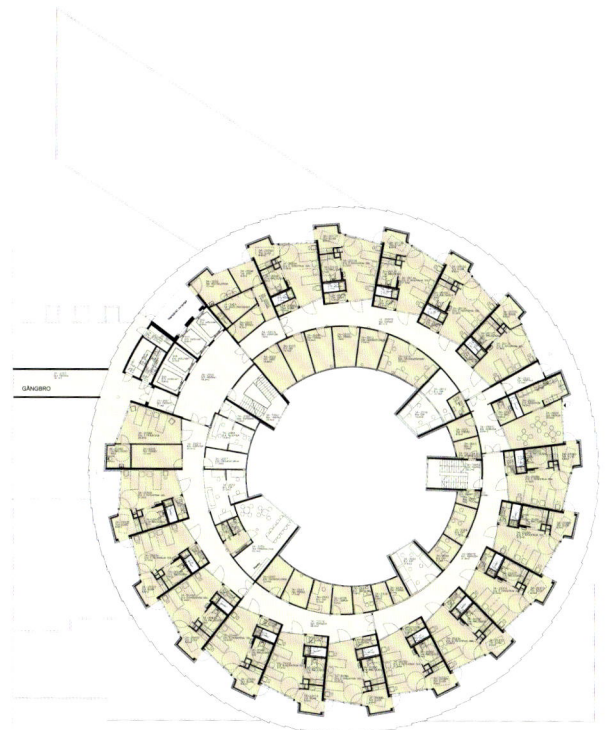

Level02-04_Infections_Ward

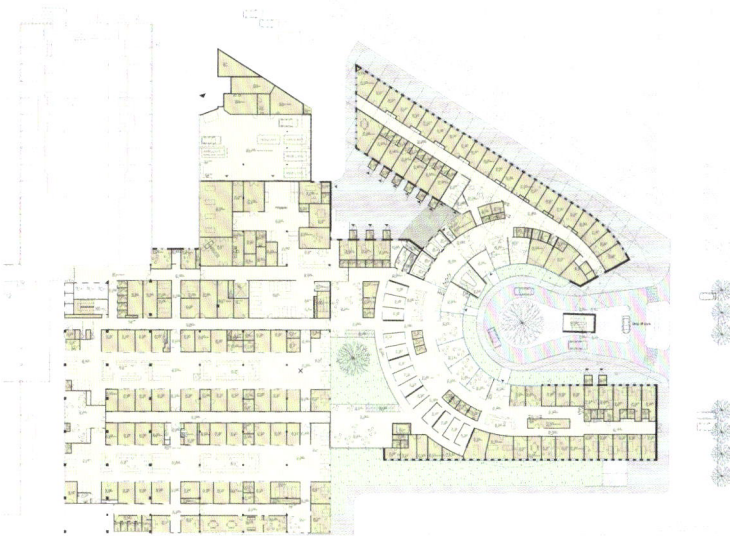

Level01_Groundfloor Access

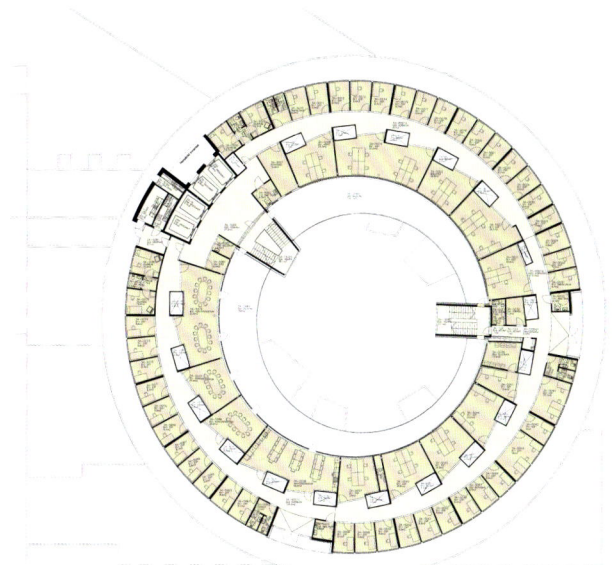

Level05-06 offices

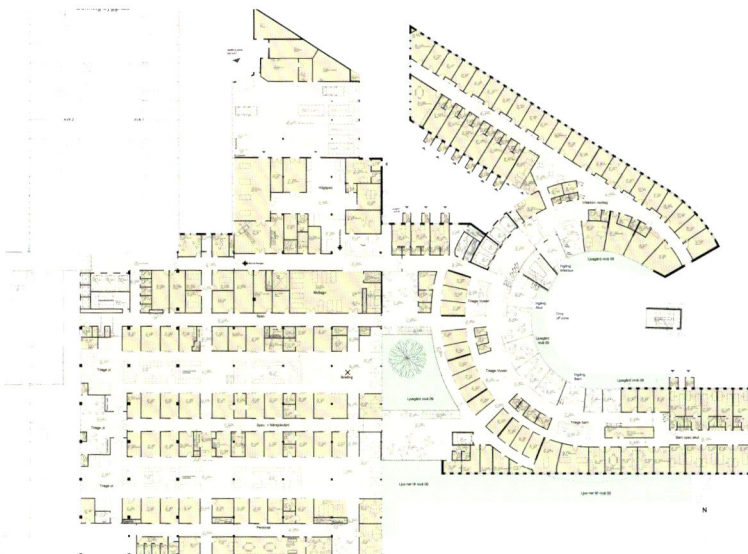

Level01_Groundfloor_Emergencies

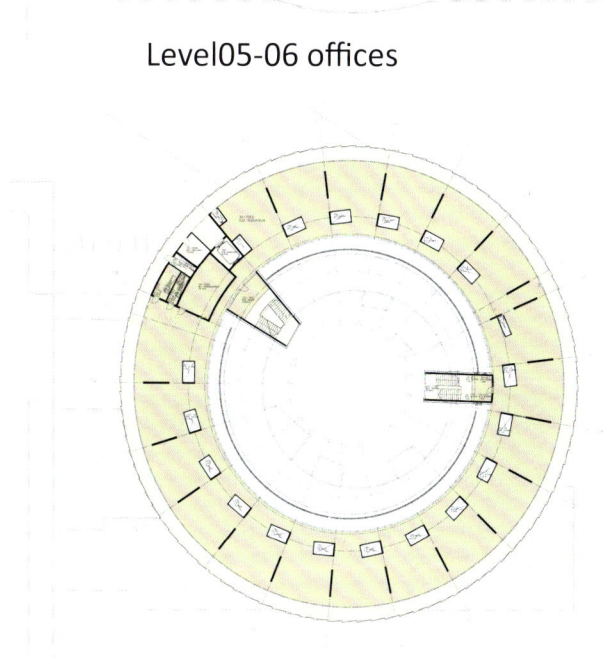

Level07 plant

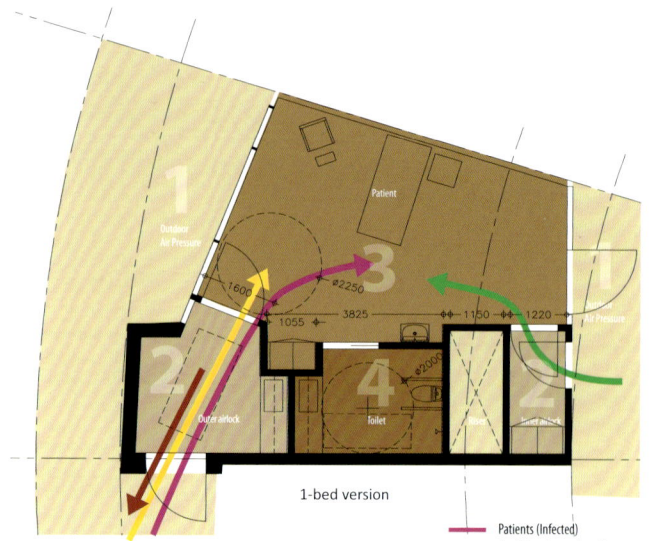

1-bed version

Patient

Outer airlock

Toilet

Patients (Infected)
Staff from internal corridor
Visitors
Waste

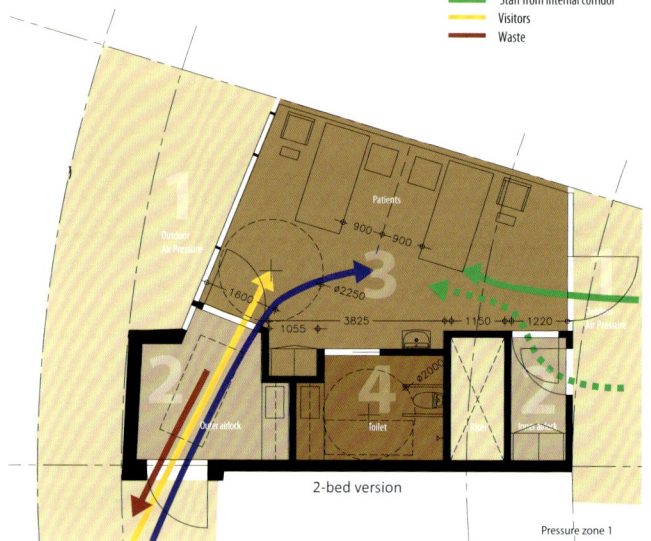

2-bed version

Patients

Outer airlock

Toilet

Pressure zone 1
Pressure zone 2
Pressure zone 3
Pressure zone 4

　　瑞典Malmö大学附属医院内的急救和传染病大楼是一个圆柱形的建筑。这座建筑的建造目的在于把疾病传染最小化。它独特的设计也让其成为医院建筑群里面的一个新地标。病人进入隔离区域后，气闸会阻止空气向外传播。外部升降机为患传染病的病人和运输医院废弃物专用，而内部升降机则为员工、输送物资和干净物品传输服务。如果发生了传染病，每一层都可以单独隔离出来。

　　此外，C. F. Møller还为急救和传染病区设计了特别设备。

Fascinating Architecture

ATP Architects and Engineers

With more than 400 employees, ATP Architects and Engineers is one of the largest integrated design offices in Europe, specialising in integrated design of complex construction projects.

As an integrated design company ATP Architects and Engineers offer "from a single source" all necessary planning design services – from the project idea to the completed building. This culture of simultaneous and interdisciplinary cooperation is highly inluenced by the fact that architects and engineers work alongside full time on speciic projects over a period of years. ATP architects lead this Integrated Design Process. Their close cooperation with ATP engineers leads to innovations in all design phases which make a major contribution to the sustainable quality of the architecture of our buildings. ATP architects and engineers invest their experience and expertise in producing innovative, resource-saving and energy-optimised buildings of a higher aesthetic quality.

C. F. Møller Architects

C. F. Møller Architects is one of Scandinavia's oldest and largest architectural practices. Our work involves a wide range of expertise that covers programme analysis, town planning, master planning, all architectural services including landscape architecture, as well as the development and design of building components.

Simplicity, clarity and unpretentiousness, the ideals that have guided our work since the practice was established in 1924, are continually re-interpreted to suit individual projects, always site-specific and based on international trends and regional characteristics.

Over the years, we have won a large number of national and international competitions. Our work has been exhibited locally as well as internationally at places like RIBA in London, the Venice Biennale, and the Danish Cultural Institute in Beijing.

Today C. F. Møller Architects has app. 275 employees. Our head office is in Århus and we have branches in Copenhagen, Aalborg, Oslo, Stockholm and London, as well as a limited company in Iceland.

Projektil Architekti

The Projektil architekti studio was founded in 2002 by Roman Brychta, Adam Halíř, Petr Lešek, and Ondřej Hofmeister. The first implemeted project was Sluňákov – Center for Ecological Activities. This building won the Grand Prix of Czech Architecture in 2007. The following project – building which was opened in September 2008 is the regional RESEARCH LIBRARY in Hradec Králové. The current implementation project is the National Technical Library in Prague which was opened in September 2009.

Projektil architects – a studio of four young architects - designs private and public buildings, urban plans, interior designs, and exhibition spaces. The architects are interested in innovations in typology and sustainable development. They invite experts, artists, scientists, and designers for cooperation on their projects.

The motto for the work of the studio is: team, openness, and interdisciplinary discussion.

Patrik Pedó and Juri Pobitzer

The architecture office monovolume has been working in the sector of architecture and design since 2003 piloting projects that go from urban design to interior design and furnishing.

The architects met at the faculty of architecture at the university of innsbruck where they have already collabo¬rated and worked together on projects. The participation at several national contests has given them the opportunity to carry out a number of successful projects which laid the foundation stone of the actual teamwork and the beginning of the mutual professional activity.

Our workplace becomes an innovative scene. How is it possible to create an innovative site and how do architecture and design react thereon? Engagement – the common object is the way to succeed. Complexity people who are faced with many impressions and opinions nowadays, need a well-linked mental activity which reduces complexity develops a sense for important factors and which is able to concentrate on synergies and contents.

Tony Owen

Tony Owen is the principle of Tony Owen Partners. A Sydney based practice with 15 staff. Tony graduated from the University of NSW with the University Medal and the Board of Architects Prize. He won the BHP Student Biennale national design prize twice in a row. Tony studied advanced architectural design at Columbia University in New York where he was awarded the graduation design prize. At Columbia University, Tony worked with some of the leaders in contemporary design philosophy including Steven Holl, Bernard Tschumi and Greg Lynn. Here he explored contemporary approaches to design focusing on complex urban issues of density, sustainability and infrastructure as well as utilising current computer and catia technology in design generation.

Tony Owen Partners is exploring new modes of sustainable design using digital technology to respond to changing environmental conditions at a micro design scale.

Stanley Saitowitz

Stanley Saitowitz / Natoma Architects Inc. is committed to design excellence. The theoretical position establishes a particular concept for each project, giving unique measure to the specific program and site. The approach to design is considered "human geography", and is especially cogniscient of the relation of building and setting. The aim is to invent spaces of passional quality.

STANLEY SAITOWITZ is PROFESSOR EMERITUS OF ARCHITECTURE at the University of California, Berkeley, and has taught at a number of schools including the GSD, Harvard University (Eliot Noyes Professor 1991/2), University of Okalahoma (Bruce Goff Professor, 1993), Southern California Institute of Architecture, UCLA, the University of Texas, the University of the Witwatersrand, Cornell, and Syracuse. He has lectured extensively in the USA and internationally.

Todd Saunders

Saunders Architecture is a firm owned by Canadian Todd Saunders who has lived and worked in Norway since 1997. Saunders has worked in countries such as Austria, Germany, Russia, and Latvia. Currently, the office is working mostly in Norway, and has projects in England, Denmark, Sweden, and Canada. The office consists of four architects from Canada, Germany and USA.

The work combines a Nordic design sensibility with environmental concerns. Each project is unique and inventive. Every project has a new process. This strategy derives from an ability to be inventive and to constantly question the purpose of our buildings. Depending upon the setting and the program, each building suggests a critique of urban planning, provides solutions to contemporary housing solutions, or creates sympathetic yet robust new forms for residential housing that are additions to the dramatic landscape in which they sit.

Adriana Sternberg & Mariano Kohen

Adamson Associates Architects is a leading architectural firm that provides design and management services to high profile clients and projects around the world. An office consolidation led to the Toronto office relocating to the historic McGregor sock factory on Wellington Street near Spadina. This preferred location on "architect's row" and close to public transit reduces transportation time and associated impacts. The historic nature of the building, with its high ceilings and exposed wooden posts and beams, allowed for an open floor plan and interesting, creative office design. The office space is 3967m² located over four floors. The fit-up scope of Adamson's space consisted of HVAC upgrades, complete electrical works including light fixtures and associated controls, architectural furnishings, finishes, and equipment.

Ilmari Lahdelma

Architects Lahdelma & Mahlamäki Ltd. was founded in 1997 in Helsinki. The partners of the company are Professor Ilmari Lahdelma, and Professor Rainer Mahlamäki, both members of the SAFA (Finnish Architects Association).

The partners are working together since 1985, having previously worked together at Architects Company 8 Studio in Tampere and at Architects Kaira-Lahdelma-Mahlamäki, Helsinki.

Our team has extensive experience in all aspects of architecture: public buildings, residential buildings, renovation projects, urban planning as well as interior architecture and furniture design. Significant part of our work has started thru architectural competitions, in which the partners have received 35 first prizes (and 47 other prizes)! The latest winning competition entries are the Kastelli School competition in Oulu and the new Embassy of Finland in Tokyo.

Perkins Eastman

Perkins Eastman is among the top architecture and design firms in the world. The firm prides itself on inventive and compassionate design that enhances the quality of the human experience. Because of its depth and range, Perkins Eastman takes on assignments from niche buildings to complex projects that enrich whole communities. The firm's practice areas include education, housing, healthcare, senior living, corporate interiors, cultural institutions, public sector facilities, retail, office buildings, and urban design. Perkins Eastman provides award-winning design through its domestic offices in New York, NY; Boston, MA; Arlington, VA; Charlotte, NC; Chicago, IL; Oakland, CA; Pittsburgh, PA; and Stamford, CT; and internationally in Dubai, UAE; Guayaquil, Ecuador; Mumbai, India; Shanghai, China; and Toronto, Canada.

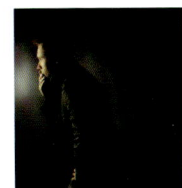

Cyrille Druar

Born 15th May 1980 in Paris
General studies until earning a Baccalaureate specialised in Fine Arts
Enters art school in 1999 -Does several internships and notably works on the renovation of Karl Lagerfeld's fashion studio and the creation of villas.
Graduates from Esag-Penninghen in 2004 at the top of his class with a degree in Interior Architecture and Design
Creation of "Cyrille Druart" company in 2007, based in Paris
Opening of I-WAY in July 2008
Designer, Interior Architect, also passionate about Photography, Cyrille Druart excels in many fields. He grew up in an artistic environment, immersed in surrealism on his mother's side, which has had a profound influence on his work and interest in the imaginary. His father, an industrial designer, exposed him to the technical side of the profession. A creator at heart, he strives to design modern, non-decorative environments, but with an underlying language and sense of wonder. Architecture thus becomes a platform for introspection.

Fjmt Studio

Fjmt offers services in architecture, landscape architecture, master planning and interior design. With a staff of more than 75 personnel, the fjmt architectural team has a wealth of architectural and technical capability to assist them as required. This includes specialists in Landscape Architecture, Interior Design, Urban Design, 3D Visualisation (all tertiary qualified) and a team of administration/non-architectural staff including a practice manager, graphic designers and IT support.

Fjmt's team of professionals are proficient in, and aided by, the latest visualisation, communication and documentation technology. The practice has placed much emphasis on technology (computer hardware and software) to facilitate the design process. Accordingly, fjmt has the latest in CAD capability; designing and documenting using a fully transitioned 3D methodology (ArchiCAD) and the most powerful computer workstations and have an array of high-speed and large format colour printers.

Juergen Mayer H.

Founded in 1996 in Berlin, Germany, J. MAYER H. focuses on works at the intersection of architecture, communication and new technology. Recent projects include the Town Hall in Ostfildern, Germany, a student center at Karlsruhe University and the redevelopment of the Plaza de la Encarnacion in Sevilla, Spain. From urban planning schemes and buildings to installation work and objects with new materials, the relationship between the human body, technology and nature form the background for a new production of space.

Juergen Mayer H. is founder and principal of this crossdisciplinairy studio. He studied at Stuttgart University, The Cooper Union and at Princeton Universtiy. His work was published and exhibited worldwide and is part of international collections like the MoMA New York and SF MoMA. His work was awarded with numerous prizes. Juergen Mayer H. taught at Princeton University, University of the Arts Berlin, Harvard University, Kunsthochschule Berlin, the Architectural Association in London and is currently teaching at Columbia University in New York.

Archimedialab
Bernd Lederle

Archimedialab is a full service architecture design laboratory aiming to combine unconventional spatial explorations with ecological and innovative technologies to create unique architectural solutions in all phases of the design and construction process.

The main focus is on concept and design and construction of cultural and public institutions, hotels and conference facilities, housing, landscape design, urban planning and commercial design worldwide.

Sternberg Kohen Arquitectos

Sternberg Kohen Arquitectos studio was created in the year 2000 by architects Mariano Kohen and Adriana Sternberg. Since then, we have developed projects that embrace several thematics: collective and individual houses, buildings, hotels, retail, offices, booths for exhibitions and interior design.

The staff is composed by a team of architects that work with dedication, providing innovative and creative solutions, to respond efficiently to our clients' requirements. We offer an integral service, solving all the stages of the project until the completion of the work.

Our professional labor has given us the experience for making prefeasibility studies, architecture projects, work direction and construction.

Our work premises are creativity in design, functionality and flexibility of the spaces, the quality of materials and constructive details and the fulfillment of the established terms and budgets, making our work our best presentation form, and the cause of a constant and sustained growth since our beginning.

Paul de Ruiter

In 1990 Paul de Ruiter (b. 1962, Rhenen, The Netherlands) graduated cum laude from the Technical University Delft. In 1992 he started his research concerning energy efficient buildings. Before founding his own office in 1994 he worked with leading architectural offices in Canada, Australia and the Netherlands. In 1994 he founded Architectenbureau Paul de Ruiter b.v., an office that is keen on research in order to design buildings and towns in which people feel comfortable and safe, without prejudicing the environment and economic feasibility.

Architectenbureau Paul de Ruiter just moved to a new sustainable office building it developed itself.

He is now, among other things, working on the TNT Green Office, one of the most sustainable buildings in the Netherlands.

Besides his design activities Paul de Ruiter works at a doctoral dissertation on climate active facades. He gives lectures, writes articles for professional journals and teaches at the Dutch technical universities.

Foster and Partners

Foster + Partners is an architectural firm based in the United Kingdom. The practice is led by its founder and Chairman, Norman Foster, and has constructed many high-profile glass-and-steel buildings.

Established by Norman Foster as Foster Associates in 1967 shortly after leaving Team 4, the firm was renamed in the 1990s to more accurately reflect the influence of the other lead architects.

It is now a worldwide practice, with project offices in more than twenty countries. Over the past four decades the company has been responsible for a strikingly wide range of work, from urban masterplans, public infrastructure, airports, civic and cultural buildings, offices and workplaces to private houses and product design. Since its inception, the practice has received 470 awards and citations for excellence and has won more than 86 international and national competitions.

Olson Kundig Architects

Olson Kundig Architects began its creative existence with architect Jim Olson, whose work in the late 1960s explored the relationship between dwellings and the landscape they inhabit in the Northwest. Olson started the firm based on some simple ideas: that buildings can serve as a bridge between nature, culture and people, and that inspiring surroundings have a positive effect on people's lives. Rick Sundberg joined the firm in 1975, and its commitment to urbanism and civic life became evident as they began designing and developing modern urban buildings in and around Seattle's national historic districts Pike Place Market and Pioneer Square.

In 1996, Tom Kundig joined Olson and Sundberg as an owner, taking the firm to another level of creative exploration and helping it grow into an office with an international reputation. Alan Maskin and Kirsten Murray joined the owners group in 2008, continuing the evolution of the firm.

CO Architects

Founded in 1986 as a regional office of San Francisco-based Anshen + Allen, we became an autonomous practice a decade later. In 2005, to acknowledge the evolution of our firm, we formally changed our name from Anshen + Allen Los Angeles to CO Architects.

Our new name is intended to evoke our collaborative approach to determining the strategic needs of each project and the coherence we seek in our built work. These fundamental aspects of our firm remain unchanged, along with the active presence of our longtime senior partners, our 85+member staff and the nature of the work we do.

We are nationally recognized experts in academic, research lab and healthcare planning and design. Throughout our history, we have maintained that technical systems, planning, documentation and project management are all integrally important in the service of design. Numerous awards attest to our reputation for elegant solutions to the demands of complex projects.

Pieta Linda

Pieta Linda Auttila (b. 1974) is an interior architect who graduated from the University of Art and Design Helsinki in the spring of 2009. Auttila has gained life experience and understanding by living in several countries and working in numerous professions around the world. Her work history includes working in a travel agency as an inquirer looking for unspoilt future tourist resorts as well as jobs in modern, experimental design agencies. The countries she has lived in include Spain, Turkey and Greece. At times Auttila has led the life of a nomad with the tribes wandering on the shores of the Black Sea. Rather than proceeding in a systematic, goal-oriented way, she has intuitively hurled herself into new projects. Auttila avoids stagnation. She wants to keep moving and make discoveries on the way. Greece is also the site of Auttila's thesis. Mensen Apartamentos is a building complex for three apartments on a high mountain ascent in southern Crete.

The thesis includes, alongside with the architecture, the material choices, solid furniture design and other out and indoors aesthetic and functional solutions. The WISA Wooden Design Hotel is Pieta Linda Auttila's first work made entirely of wood. In the competition she was especially attracted by the variations and possibilities of one material.

Daniel Libeskind

Daniel Libeskind is the Principal Design Architect in Studio Daniel Libeskind and solely responsible for all design decisions. Every project is developed with a consistent core team, which works together throughout the design process toward the eventual realization of the project. Within the Studio, teams are set up in individual project rooms and great emphasis is placed on working together.

The proximity of the different team members ensures communication on a daily design development basis. Studio Daniel Libeskind operates with thoroughly integrated teams, where team members assume a high degree of responsibility for ensuring that the project management function is there to support and serve the primary creative design function. There is a clear reporting structure within the team. Design and planning issues are sketched by Mr. Libeskind, alternatives drawn up by the team, discussed and reworked with Mr. Libeskind through regular internal presentations which act as in-house critiques and then reworked until the solution is agreed upon by him. The Project Architects coordinate the work of the team and maintain continual design contact and development with Daniel Libeskind ensuring the ongoing quality of the work in the well-recognized Studio tradition.

Antonio Vaíllo i Daniel - Juan L. Irigaray Huarte

ANTONIO VAILLO i DANIEL, ARCHITECT, (BARCELONA, 1960)
1979-1985   Architectural Studies in ETSA University of Navarra.
JUAN LUIS IRIGARAY HUARTE, ARCHITECT, (NAVARRA, 1956)
1974-1980   Architectural Studies in ETSA University of Navarra.
COMPETITIONS AND PRIZES (only list some of them)
2010
First Prize. Archizinc Awards. B3 House. Pamplona
First Prize. COAVN Awards. Lounge MS. Cadreita
First Prize. Restricted competition
2009
Gold Medal. Miami Beach 2009 Bienal. Interior Design. D Jewellery Shop. Pamplona
First Prize. Restricted competition
First Prize. Interior FAD'09 Opinion Award. Elmerca'o Restaurant. Pamplona
TOP 50 Interior of Spain 2009. Convened by VIA Group
2008
First Prize. Restricted competition
First Prize. Interior FAD'08

TWS & Partners

Kim Herforth

Simone Micheli

TWS & Partners was established in 1998 in Indonesia. It is led by Tonny Wirawan Suriadjaja.

Tonny Wirawan Suriadjaja was born in 30 January 1972, Jakarta, Indonesia, is an architect who always tries to find any innovation in architectural and design interior.

May 1995 - July 1995 worked with Gunawan Tjahjono Ph.d to design National Museum in Korea

July 1995 - November 1995 worked with Shimizu Lampiri Consultant at Jakarta as an Architect

November 1995 - June 1998 worked with Ciputra Development at Jakarta as an Architect

June 1998 - Present established TWS & Partners Design Team

3XN was founded as Nielsen, Nielsen and Nielsen in Aarhus in 1986 by the architects Kim Herforth Nielsen, Lars Frank Nielsen (partner until 2002) and Hans Peter Svendler Nielsen (partner until 1992). The three Nielsen architects, often referred to as the Nielsens – and today simply as 3XN – quickly became known for two things: their preference for ground-breaking architecture, in defiance of the anti-humanistic modernism, and projects demanding a high level of detail and employing workmanship of the highest quality.

3XN became a limited liability company in 1995. The firm comprises of two offices each with a staff of 60 – one in Kystvejen in Aarhus, and one in Strandgade in Copenhagen. We work, however, as one practice continuously interchanging ideas and staff between the two offices. We operate in a project-oriented network structure, and at every stage we team up in workshops with partners, experts, users etc. At the workshop we discuss the results of our research, develop the concept and a framework for the design work.

Simone Micheli is a university contract professor. He founded the architectural studio with his name in 1990 and the design company "Simone Micheli Architectural Hero" in 2003. His works concerning architecture, contract, interior design, exhibit design, design, graphic and communication are strictly linked to the sensorial glorification. He is the curator of some experimental events for some of the most qualified international fairs. He shows his projects in the most important worldwide architecture and design exhibitions. He is the curator of some thematic "contract" exhibitions in the most important international fairs of this field. He represented Italian interior design taking part to the XXX Congreso Colombiano de Arquitectura in Baranquilla and participating at the International Conference of Architecture in Hanover; in 2008 he has signed the event called "La casa Italiana" for ICE and Fiera Verona in collaboration with Abitare il Tempo - Acropoli, at the MuBE (Brazilian Museum of Sculpture) in San Paolo.

Concrete Architectural Associates

Goedele Desmet, Ivo Vanhamme, Jean-Michel Culas

Doriana & Massimiliano Fuksas

Concrete Architectural Associates is founded in 1997 by Rob Wagemans, Gilian Schrofer and Erik van Dillen. They met each other by a not realized project, a head office in Amsterdam for Cirque du Soleil.

Rob Wagemans (present director), born in Eindhoven on 13th of February 1973. History of education: Master of Architecture, Academy of Architecture in Amsterdam.

Erik van Dillen (at this moment he is just creatively involved with Concrete), born in de Bilt on 27th of April 1960. History of education: interior architect, Rietveld academy, catering industry skills in the kitchen, painting restorer.

Gilian Schrofer left concrete in 2004 to start his own company.

Concrete Reinforced is founded in 2006 by Rob Wagemans and Erikjan Vermeulen. Reinforced is responsible for architecture and urban landscaping and enriches concrete with its disciplines.

Erikjan Vermeulen (present co-director), born in Texel on the 2nd of April 1973. History of education: Master of Architecture, Academy of Architecture in Amsterdam. Worked for different architects to start his own company in 2003.

Architecten Bob361 architectes, is an architecture office run by Goedele Desmet, Ivo Vanhamme and Jean-Michel Culas. There are 2 offices. One in Brussel and one in Paris.

Awards

2006 Leuven, 17 social houses Heidebergstraat, Belgian Building, Awards, Nomination

2005 Borgerhout, Hydraulic Laboratorium, Welstandsprijs, Nomination

2004 Borgerhout, Hydraulic Laboratorium, Steel Construction Award, Award for a Renovation Project

2004 Hydraulic Laboratorium, Borgerhout, Prix Georges Hens de l' Academie Royal de Belgique, nomination

2001 Linden, addition to the CLP house, Renospecto Architectural Award, Nomination

2000 Lebbeke, 4 patio-houses, European Award Luigi Cosenza, Special Mention

1999 Lebbeke, 4 patio-houses, Award of the Belgian Architecture

1998 Leuven, addition to the MC house, Architectural Award of the City of Leuven

1996 Paris, renovation Cavalerie apartment, Architectural Awards, Special Mention

Native of Lituania, Massimiliano Fuksas was born in Rome in 1944, where he graduated in Architecture at "La Sapienza" University in 1969.

In 1967 he created his own studio in Rome, followed by a second one in Paris in 1989. He opened an office in Vienna in 1993, and in Frankfurt in 2002, active until 2001 and 2009. From 2008 he has an office also in Shenzhen, China.

He was Visiting Professor at several universities.

For many years he has been devoting special attentions into urban problems in large metropolitan areas.

Doriana O. Mandrelli was born in Rome where she graduated in History of Modern and Contemporary Architecture in 1979 at "La Sapienza" University.

Since 1985 she has been cooperating with Massimiliano Fuksas. Since 1997 she has been responsible for Fuksas Design.

She was the curator of the Architecture's Section at the VII Biennial of Architecture in Venice (2000), she was in charge of four special branches. Her projects are characterized by a continuous research in new materials and innovative technology.

They work and live between Rome and Paris.

Endo Shuhei

PROFILE

1960 Born in Japan

1986 Obtained a master's degree at Kyoto City University of Art

1988 Established Shuhei Endo Architect Institute

2004 Professor at Salzbulg Summer Academy

Currently professor at Graduate School of Kobe University

AWARD

2000 Third Millennium International competition of ideas la Biennale di Venezia, 1st prix, Italy

ar+d award 2000, Grand Prix, England

2001 Premio Internazionale di Architettura Francesco Borromini , Italy

2002 8th Public Architecture Award, Japan

2003 Kenneth F. Brown Asia Pacific Culture and Architecture Design Award, Grand Prix, U.S.A

Education Minister's Art Encouragement Prize for Freshmans, Japan

2004 9th International Architecture Exhibition Special Award la Biennale di Venezia 2004, Italy

2005 Architecture and Design for Children International Award 2005, UK

BSC Awards, Japan

2006 Annual Architectural Design Commendation, Japan

2007 Osaka Architectural Award, Japan

2008 ARCASIA Award Gold medal, Sri Lanka

2009 Green Good Design Award, Chicago

2009 ARCHIP Architectural Award, Moscow

后 记

    本书的编写离不开各位设计师和摄影师的帮助，正是有了他们专业而负责的工作态度，才有了本书的顺利出版。参与本书编写的人员有：Antonio Vaillo i Daniel, Juan Luis Irigaray Huarte, David Eguinoa, Bernd Lederle, Wolfgang Heckmann, Tina Kierzek, Tilo Weber, Katharina Schneider, Jonas Beer, Achim Zumpfe, Paul de Ruiter, Patrik Pedó, Juri Pobitzer, C. F. Møller Architects, Cyrille Druar, Daniel Libeskind, Richard Francis-Jones, Stanley Saitowitz, Projektil Architekti, Doriana & Massimiliano Fuksas, Goedele Desmet, Ivo Vanhamme, Jean-Michel Culas, Jim Olson, Stephen Yamada-Heidner, John Kennedy, Olivier Landa, William Franklin, Megan Zilmmerman, Michael Picard, Debbie Kennedy, Cristina Acevedo, Scott Kelsey, Jorge de la Cal, Fabian Kremkus, Neil Kaye, Markus Bischoff, John Winder, Derrick Chan, Adamson Associate Architects, Andrei Florian, Juergen Mayer H., MarcusBlum, Jan-Christoph Stockebrand, Wojciech Witek, Magdalena Skoplak-Seweryn, Jakub Kaczmarczyk, Pieta Linda, TWS & Partners, Simone Micheli, Rob Wagemans, Erikjan Vermeulen, Jeroen Vester, Sander Vredeveld, Matthijs Hombergen, Julian Weyer, Kessler & Kremer, Michael van Osten, Kim Herforth, Perkins Eastman, Foster and Partners, Nigel Dancey, Iwan Jones, Dominik Hauser, Rachel MacIntyre, Adrian Nicholas, Declan Sharkey, Stefan Unnewehr, Pietjy Witt, Endo Shuhei, Tony Owen, Roman Brychta, Adam Halír, Petr Lešek, Ondrej Hofmeister, Ilmari Lahdelma, Todd Saunders, SAMARK Arkitektur & Design., Jose M. Cutillas, Pedro Pegenaute, Daniel Galar, Oskar Riz, Bitter Bredt Fotografie, Michele Nastasi, John Gollings, Andrew Chung, Mathieu Faliu等。